An Elk

Beverly Lein

Library and Archives Canada Cataloguing in Publication
Lein, Beverly, 1949-
An elk in the house / by Beverly Lein.

ISBN-13: 978-1-896300-99-3
ISBN-10: 1-896300-99-5

1. Elk—Alberta—Manning—Biography. 2. Lein family.
3. Elk farming—Alberta—Manning. I. Title.

SF459.D4L44 2006 636.2'94 C2005-907693-3

Board editor: Don Kerr
Cover design: Tobyn Manthorpe
Author photo and interior images: Courtesy of the author.

Every effort has been made to obtain permissions for photographs. If there is an omission or error the author and publisher would be grateful to be so informed.

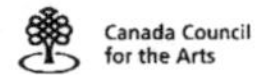

NeWest Press acknowledges the support of the Canada Council for the Arts and the Alberta Foundation for the Arts, and the Edmonton Arts Council for our publishing program. We also acknowledge the financial support of the Government of Canada through the Book Publishing Industry Development Program (BPIDP) for our publishing activities.

NeWest Press
201–8540–109 Street
Edmonton, Alberta T6G 1E6
(780) 432-9427
www.newestpress.com

NeWest Press is committed to protecting the environment and to the responsible use of natural resources. This book is printed on 100% post-consumer recycled and ancient-forest-friendly paper. For more information please visit www.oldgrowthfree.com.

1 2 3 4 5 09 08 07 06

PRINTED AND BOUND IN CANADA

To my editor Dianne Smyth whose faith in what I could do or what I could write never wavered. Every new author should be blessed with an editor like Dianne. Her knowledge and wisdom are invaluable.

To the NeWest Press Board who took the time and patience to read the manuscript and send the story on its way.

To my grandchildren Brittany, Morgan, Sydnee, Rachel and Ashley who believe Granma can do anything until she sits down and cries, then they swarm around me like little bees to console and protect me.

To my husband Carson, son Darren, and his wife Alison and daughter Lana, who always shook their heads at me but never tried to stop me.

To my sister Louise who has had to listen to every word and story I have ever written.

To my brothers-in-law Alvin and Curtis, and friends Darrel Hunter, Dwayne Veldhouse, Jim Gordey, and other friends that helped in times of need, I thank you all.

—Beverly Lein

PART 1
An Elk in the House

The cold chilled my bones as I drove out to the elk herd to do chores on that grey, rainy morning. I was tired. The middle of calving season is very stressful. And even if things do go well, I worry.

As I drove to the pasture I thought about how much I wished our herd was out of town, away from the highway traffic. Our elk farm nestles on the outskirts of the little town of Manning, Alberta. It contains about twelve hundred people, give or take a birth or death.

Because the farm runs along Seventh Avenue we named the ranch Seventh Avenue Elk Ranch. I had really wanted to name it after our deputy mayor, Del Harbourne, who was a big supporter of the ranch when we started up. But we couldn't come up with a name that sounded right. Del, heart and soul, has been dedicated to our little town as far back as I can remember.

Yellow 53—The Heifer who Couldn't Stand Up

All these thoughts were passing through my head as I started lifting pails from the back of the pickup truck, the mud making it more difficult than usual. I was trying to hurry before the herd got to me. It's hard to dump the oats and watch your back at the same time, especially important during calving season. The cows are dangerous at this time of the year because they are protecting their calves, which makes it very difficult to tag a calf—something we do the day after it is born. If a calf squeals you have the whole herd coming down on you. I mean the whole lot of them, even the ones that don't have a calf. I have always said that humans could take a lesson on raising children by watching a family of elk. Their love and devotion is amazing to see.

As I stood shivering in the rain, watching the mothers feeding and nurturing their baby elk, I was wishing I had put a sweater on under my coat. I turned to get into the truck and noticed a mother cow trying to help her calf. The calf would start to get up, then fall on her little head. I got the binoculars out of the truck to get a better look and saw that she had a yellow tag in her ear, Number 53. I looked at the book I keep in the truck with the birth dates and sex of each calf and saw that Number 53 was a little heifer (female) calf born on 30 May 1999. As I stood watching the calf, I realized she was in trouble. I had no idea then what an important part of our lives Yellow Number 53 would become.

Doc Dewey's Initiation into the Lein Family

I went to the house and phoned my husband,

Carson, who is a truck driver and farmer—actually he's a jack of all trades, which men of the north country have to be if we are to survive. I then called Dr. Dewey Stickney, our local veterinarian, whose clinic is right across from us. Carson and Dewey arrived at the house around the same time. I watched Dr. Stickney walking to the house. Neither of us could have known that, from that moment on, he would be adopted into the Lein family. Sometimes people join a family whether they want to or not, and this is what happened to Dr. Stickney (who soon became Dr. Dewey, also known as "Doc") and his staff. We had no idea how interesting our lives were about to become as we drove to the pasture to look at the little heifer who couldn't stand up.

Dr. Stickney parked his noisy, red, diesel pickup close to the calf. The mother elk, Blue 72, ran a little ways away as we got out, but not for long. As Dewey examined the calf, Carson and I kept the mother off of him. As I mentioned before, elk mothers are very protective and can be extremely dangerous. Blue 72 was no exception as she tried to run us off. Dewey's quick diagnosis was that the calf had temporary paralysis in her right front shoulder and leg. She hadn't sucked and was badly dehydrated. She needed colostrum quickly; colostrum is the first milk produced by the cow after the birth of a calf. She was already two days old and it was amazing that she was still alive. It is every elk rancher's nightmare to have to bottle-feed an elk calf. There are always complications and it is a lot of hard work.

Carson thought we should put her down right away and save ourselves the pain of losing her in a couple of days. He knows how emotionally involved I get with these little critters. I have always believed that lots of love and hard work fixes everything, with both children and animals. Unfortunately, this isn't always the case.

Where There's Life There's Hope

Dewey looked at me and said, "Your call Bev. What do you want to do? She's in pretty rough shape and her chances aren't very good, but it's up to you." My next thought was, "How come when there is a hard decision to make, I have to make it?" On the other hand, where there is life there is hope, so I told the men we were taking her in. Dewey has an ironic, dry sense of humour. He reminded me every chance he got over the next couple of years of trouble and sorrow that it was my decision to save her.

As we picked up the calf, she started to cry for her mother, who chased us as we scrambled for the pickup. Carson threw her into my lap and jumped into the back of the truck. Dewey dove into the driver's seat and slammed his door as the mother roared around to his side. As I sat trembling, trying to hold on to the calf while she fought to get away from me, I wondered if I was doing the right thing.

Blue 72 chased the pickup all the way to the gate, calling for her calf. I felt so sad for her as she stood there with her swollen milk bag. She could not understand why she was losing her calf. When we got to the clinic, Gladys Fazikos, an animal health technologist and Dewey's right arm, mixed some colostrum while Bev Warren made a bed in a cage for the calf. I asked why they were putting her into a cage, and Gladys explained, "So she doesn't throw herself around and hurt herself. The less she struggles, the better off she'll be. She's so weak; we'd like her to save the strength she has."

It's Touch and Go

For the next few days it was touch and go. I tried to get her to suck on a bottle. After fifteen minutes of failing, Dewey agreed to tube her. This is a procedure used to get milk into calves that refuse to suck. A feeding tube is run through the calf's mouth down into her stomach. We fed her a cup of colostrum every two hours. Before each tubing, Dewey had me try to give her the bottle. Both the calf and I hated the tubing, but she would not wrap her tongue around the bottle nipple and suck. We didn't get much rest with feedings every two hours, but it was necessary, as a baby elk has to be stimulated in order to learn to suck. I had to do all the cleaning with water and a cloth, while a mother elk does it with her tongue. A mother elk will spend hours licking, stimulating, and feeding a newborn at this age.

Because the calf was not able to stand, I was becoming exhausted. I was on my knees, on the ground, up and down until every bone in my body ached. At times I cried from pure discouragement. In frustration I would say, "You little dummy! Suck! Why won't you suck?" After each feeding I would pick her up and try to make her stand. I'd reach across and straighten her leg out to get her to walk until I thought my back was going to break. Carson helped when he could. When I was played out, Gladys and Bev would try for a while. One little calf was exhausting all of us! Still, I kept telling myself not to get attached, not to dare to love her. I knew I would probably lose her in the end, and it was important that I stay tough.

Dewey was a great help. He agreed to leave the side door of the clinic open so I could come and go as I pleased. On the second night my phone rang. It was the clinic. I was sure they were phoning to tell me the calf had died, but Dewey was just calling to tell me to get some sleep. He was going to tube the calf so I could get some rest before I fell over from exhaustion. He was a great one to talk! It seemed to me that he and his staff were on the go day and night without food or sleep.

I hurried to the clinic the next morning, anxious to see if my precious calf was still alive. She was badly dehydrated, but her vital signs were getting better. I hurried home to do the chores and check the herd to make sure the other elk were okay. We had four more calves born over a three-day span.

Blue 72 Finds a New Baby

I checked on Blue 72. I knew she was sad and I could still hear her calling her calf. She was constantly standing or lying down in the spot where we had taken her baby. That day another cow was having a calf a few feet away from Blue 72. When the calf was born, Blue 72 went over to the new mother, Yellow 74, and started to eat the afterbirth and help her clean up. I noticed that she licked the other cow's calf just like it was her own. Amazingly, the other mother elk was letting her do it. I wondered what these animals would do next! Yellow 74 let Blue 72 move right in, and together they shared this little calf, both letting the baby nurse from them. Blue 72 quit calling for her calf that day. She thought she had found her.

I was relieved that everything was fine in the pasture and hurried back to the clinic. It seemed that when I was with the herd I was worrying about the calf, and when I was with the sick calf I was worrying about the herd. At times I thought "Nuts" should be my name.

Yellow 53 Gets a Name—And a Mother

By the time I got back to the clinic for the next feeding, Gladys had a bottle ready. I was dreading it because the little one fought the bottle so hard. She hated it, but she hated the tubing more. I was still trying to hold my heart at a distance while going through the motions of mothering this calf when the inevitable happened. My daughter-in-law, Alison, drove up with my two-year-old grandson, Morgan, who was just learning to talk. They got out of the truck and Morgan waddled over to me where I was sitting on the ground. He said, "Better, better," but I thought he was saying, "Butter, Butter." He was trying to ask if the elk calf was better, and just like that, Yellow 53 had a name—Butter.

As little Morgan and I started to laugh, I started to cry too. Butter must have sensed our feelings for her because at that moment she forgot her mother—and bonded with me. A miracle had happened. She didn't suck yet, but for the first time she started to lap at the nipple with her tongue and swallow. The milk was finally going down her throat and not onto the ground. What a moment of elation! I had just become the mother of a baby elk named Butter. I forgot all about being tough and just loved her with my whole heart.

Butter Learns to Walk

The tubing would be stopped as soon as she was getting enough milk. We carried on the whole day, but still no sucking. Then—at four in the morning on the fourth day—Butter sucked. Of course I had to phone the whole countryside to tell them Butter could now suck her bottle! We never looked back in this department. Butter still loves to suck!

Each day brought something new. One day, Butter would have diarrhea, the next she'd have a fever. Sometimes she became listless. This calf had me scared to death with the thought that I might lose her. I never gave up. Not only did I love her, she was worth twelve thousand dollars. At each feeding I would stand her up to get her to use her leg and she would fall down. I would pick her up and start over, straighten her leg out, and make her walk. In the third week Butter realized she could walk on three legs, and eventually put weight on the bad leg. By the fourth week, Butter was walking. She still limped badly, but could get up and down on her own. That whole month my family had been telling Dewey we thought Butter needed a splint on her leg, and he kept giving us a flat "No." He told us we needed to continue to make her use the leg. He was right, as usual. Dewey explained that her positioning in the womb or trauma during birth caused her paralysis, and the only cure was to use it. The more she used it, the stronger the leg would get.

Butter Joins the Lein Family—Big Time

The minute Butter was able to suck, I brought her home. The next question was, "Where should I put her?" Carson made her a pen in the backyard, but she was always calling me. I wasn't getting any work done

and I didn't like going out in the dark at night to feed her, so we took her into the garage. That worked out well. With both Butter and our grandchildren—four-year-old Brittany, two-year-old Morgan, and one-year-old Sydnee—demanding my attention, it was easier to have them all in the house with me.

Although the idea was to have Butter stay in the garage, she refused. She was so used to being with me that she wouldn't let me out of her sight. I started sleeping on the couch with Butter on a blanket on the floor beside me. I quickly learned that when she woke during the night before feeding time, it meant she had to go out to pee. If I didn't get her outside right away, the blanket would get soaked. Because she had to have a nice dry blanket, the washing machine was going day and night until I learned to move more quickly.

Once Butter was fed and washed, she would lay down on her blanket. I would hang my arm down toward the floor, where she blissfully sucked on my finger until she fell asleep. Sometimes she would wake up for no reason and look for my finger. If it wasn't there she would cry until I put my finger down to her. Only then could both of us sleep as she sucked away.

Because this was a day-and-night ordeal, I was often very tired. One night I got up to fix her bottle. As I waited for it to warm up in hot water, I put my head down on the counter, leaning on my arms. I could hear Butter clip-clopping across the hardwood floor coming toward me. She put her little chin on the counter beside mine, watching and waiting for her bottle. I had to smile at her, she looked so cute.

Clip-Clop and a Trip in the Bathtub

The one time Butter fell on the hardwood floor, she looked just like Bambi when he fell on the ice. There she was, lying spread-eagled, her long legs straight out around her, with a very frightened look on her face. I helped her scramble back up, watching out for her sharp little hooves. She never let that happen again. Afterward she walked very carefully. Her going from not being able to walk at all to walking on a slippery surface was amazing. But I was learning that—with Butter—all things were possible. She sucked on my finger to regain her composure and kept looking down at that scary floor as she tucked herself against me.

There are three stairs from the garage up to our house. Butter had no trouble coming up, but I was petrified she'd break one of her little legs going down, so I always carried her. As she got bigger it got harder to handle her, but, again, Butter figured it out for herself. She was always in a hurry to get to me, so she'd jump with her long legs, hit the rug at the bottom of the steps (which would stop her from sliding), and she was away. Up and down she would go. Sometimes she'd just play on the steps like it was some sort of game.

One night I was awakened by Butter's calls. I heard Carson talking to her. She must have become disoriented in the dark and couldn't find me. She was down in our bedroom looking for me. Carson hollered at me to call her, so I did, "Butter, Butter, I'm out here, what are you doing? Come here!" I should have turned on the lights. I heard her coming, but she made a wrong turn and ended up in the bathroom. As she turned around, she pushed the door shut and stepped into the tub just behind the door. What a performance we had then! The bathroom is very small and things were flying all over by the time I got to her, but Carson and I had a good laugh. When things finally calmed down, I laughed until I cried. She looked so funny when I turned the lights on, sitting on her haunches in the tub! If you think it would be difficult to stand on hardwood with those little hooves, just imagine how tough it would be on enamel! Butter sure didn't think it was funny. She was very frightened and wouldn't suck her bottle, she only wanted my finger. She sucked for an hour before she finally went to sleep.

Goat's Milk Works!

When Butter was one month old, I started giving her eight ounces of milk every four hours. I then increased the amount of milk until she was happy going six hours between feedings. Butter, of course, never made anything easy for me. Regardless of which powdered milk we tried, she was always sick with diarrhea. Finally we found Tony and Linke, who lived in Hawkhills, some thirty miles away. They owned milking goats and supplied us for the next two months with nourishing, wonderful goat's milk. After just two days of goat's milk, she was never sick again. But it soon became very costly, with the gas travelling to and from Hawkhills and the need for about four litres of milk every twenty-four hours. Actually, at the peak of Butter's feeding, she would drink four litres of milk at each feeding. Add on the veterinarian bills, and we often wondered if she was worth all the expense. Of course she was—she was Butter—and she had wormed her way into our hearts.

When I went outside to work Butter was always by my side. When she was tired she would lie down for a while where she could keep an eye on me; true to the nature of an elk calf, she would find a place in the deep grass in the trees and sleep until I called her. If it was time to eat and I was late, she would stand up and call for me.

Elk! Elk!

One day, when Butter was one month old, she and Morgan were walking side by side when they both decided to go the same way. Both baby boy and baby elk fell down. All I could see were elk legs as my little grandson was getting stepped on and kicked by Butter, who was trying to get away from him. She was squealing and Morgan was crying and screaming, "Elk, elk!" I was horrified. As they regained their footing, both came running to me, my poor little grandson thought he had gotten the worst of it with all those legs stepping on him, and Butter thought she had because she couldn't get away from him. I stood there comforting my grandson in one arm until he stopped crying, with the other arm around my baby elk while she sucked on my finger and got her wits about her. Needless to say, Butter and Morgan were both pretty leery of each other for the next few days.

Butter loved watching TV with us as she lay on her blanket by the couch. When she stared at the TV screen, I swear she was actually watching it. One night Carson lay down on my couch while I used his. Butter got up and started walking back and forth, driving us crazy. She would give a little squeal, go to the blanket, and then come back to me. It took us a while to realize that Butter felt we weren't where we were supposed to be, so we exchanged couches. Butter lay down immediately. Afterward, we often played this game with her—deliberately lying on the wrong couch just to see her have a little fit.

She Doesn't Know She's an Elk

Butter was beginning to become a nuisance in the house. She would pick up cups, pictures, and ornaments with her teeth and walk around with them in her mouth. Of course, she would drop and break things as she played her little games, so, when she was three months old, much to her dismay, we moved her out of the house and took her to the elk herd. While she had always been exposed to the herd, she never got over being frightened of them. When they came close to her she'd squeal and run to me.

Carson always said, "Mom, you have to get a mirror for Butter and show her what she looks like. She doesn't know she's an elk; she thinks she looks like us." That is the saddest part about raising baby elk. While they don't belong in our world, they don't belong in theirs either. Grandpa Bud says elk don't speak the elk language when humans raise them, and that makes it tough on them.

When the herd came close, Butter would start crying for me and I'd run out to her. If I got there too late and the herd was into a feed, she would be too afraid to come to me. I was afraid to go in, so I'd call Carson to help me. He'd go in and the herd would move away a little. Butter would run between the fence and Carson with her eyes closed to get to me. I guess she hoped they'd disappear while her eyes were closed.

Butter knew Carson was her protector from the time she was little and she always minded him. Carson was never afraid, but as she got bigger, I became a little nervous of my four-legged baby. Not that she ever hurt me. Outside of bunting me for her bottle she was always gentle with me. Anyone who has raised elk has told us to never trust them, especially when they have been bottle-fed, because they have no fear of us. I always keep that in the back of my mind. I know it's true, but I would feel very guilty to think that of Butter.

Learning How to Be an Elk's Mother

As humans, we never fully comprehend all the duties a mother elk performs. We just take for granted everything they do while raising their young. In order for me to raise Butter, I had to watch the mother elk carefully and take my cue from them. Every day in the summer, twice a day, the herd moves down to the big dugout for a drink and a swim. Elk love water. The way the babies run in and out of the water, it is easy to see that they are natural-born swimmers. So twice a day I took Butter down to the dugout for a drink and a swim. Many hours were spent at the dugout as she learned what to do and what not to do.

At times she had to learn her lesson the hard way. The ends of the dugout slope in, deepening gradually, but the sides of the pond drop straight down nearly fourteen feet deep. Butter, who was about two-and-a-half months old at the time, was running up and down on the sides of the dugout having a wonderful time. She would run in on the ends of the pond, stand in the water up to her little tummy, buck around and run out, splashing both of us. She just loved it.

One time, I went for a walk alongside the dugout while she played on the bank near the water. When Butter saw me on the other side of the bank she came roaring toward me. She noticed the water but never stopped her run. She just flew through the air into the deepest part of the dugout. Down she went—the water closing right over top of her head. I was sure she was going to drown. I thought I would have to jump in and save her. Soon, though, up she came, swimming frantically away from me but—at the same time—crying and calling for me. I was screaming her name and when she heard my voice she turned her head toward me. Her head then acted like a rudder and around she came, heading back to shore, swimming as hard as she could. I helped haul her up on the bank, where she lay scared senseless. Of course she had to suck my finger until she got her composure back. Never again did Butter jump off the side of the dugout. Instead, she only played at the ends where she could control the situation.

Many summers I had watched the mother elk during storms when the thunder and lightning frighten the calves. The mothers run and gather the calves and take them to the shelter of the trees. If it happens to hail, they run to the haystack for protection. I think this is very smart, as a bad enough hailstorm can kill the calves. So, of course, my turn came with my baby when a terrible lightning and thunderstorm hit us. The rain was coming down in buckets and water was running through the yard like a flood. I immediately wondered where Butter was. I looked out the window and there was my poor baby running along the fence squealing for me. I grabbed my coat and ran to the fence to let her out. She ran toward me crying, so I headed for the thickest trees by the fence with Butter on my heels. We huddled down together beneath the thick covering of bush. I was soaking wet, miserable and muddy, but baby Butter was content, dry, and happy, lying down beside me, sucking away on my finger, contentedly watching the storm. With a few thunderstorms behind her, Butter learned to run to the trees or the haystack on her own.

Butter Gets Bigger—And Bigger!

Elk calves are usually weaned by elk farmers at three months so, oh happy day, I took Butter's bottle away from her when she was three months old. I continued only with her night bottle, which I gave to her until she was six months old. No matter what time I got home at night, Butter was always at the gate waiting for her night bottle. I knew she was spoiled, but she was special—she was my baby and I was all she had.

When Butter was about four months old, we stopped the grandchildren from going inside the fence with her. While the children stayed the same size, Butter was quickly growing and had become much too big to play with them. She loved the kids dearly, but would playfully paw at anything smaller than her and gave the kids quite a few knocks on the head with her front hooves. To this day, when the kids are outside playing, she stands by the fence hoping they will come and talk to her.

One day I heard the kids crying and ran outside to see them right by the elk gate. I called to ask if they were on the other side of the fence with Butter. Brittany told me Morgan had gone in. I had to get him out. Morgan was too young to open the gate, but Brittany—the great and wise older sister—had opened the latch. They both learned a lesson about their four-legged friend playing too rough. To be on the safe side, Grandma started locking the gate.

Butter Is an Outcast

As life went on, Butter grew, but she was not accepted in the elk herd. Adam, the bull, was the only one who would sometimes come to keep her company. But he also sensed she was different and eventually left her to herself. When springtime came and the new calves were born, Butter was very excited because she was accepted by the others as a nursemaid. She was designated as a babysitter and was allowed to lie with the calves. The cows still didn't want her around them, but this was okay with Butter because she loved the new, little babies.

I would often go out to the pasture to check on her, always hoping the others had accepted her and weren't bullying her around too badly. I was constantly watching for her tag to see where she was, but as the outcast she was always off by herself or with the calves. I was disappointed for her, and sometimes

I would cry for my lonely little elk. I wanted so badly for another elk to love Butter.

When the day was done and it was time to sleep, Butter always came back to the facility to be close to the house and to me. Come morning, when I went to do the chores, Butter would hear the truck, come running, and follow me everywhere. Whatever I did—from searching for a new calf to feeding—Butter stayed beside me. She followed the truck like a dog, and was sometimes a nuisance. If we were tagging a calf she'd have to be right in there with her head down beside Carson's. If we were standing and talking she'd have to stand in the circle and listen to the discussion as if she understood. I remember once

I had forgotten something at the house. When I drove out of the gate Butter came out with me. Now what could I do? I drove around the yard with her chasing the truck, then drove back into the pasture. She followed me right back in. Even now, I believe that if I let Butter out of the pen, she would come directly to the house and want inside.

The Facility Becomes "Home"

That fall we drove the herd into the facility, which consists of about three acres. It has cross fences with gates, so that we can run cows into bigger or smaller holding areas. We hold the cows all together in the large part of the facility, then open a gate that goes to an alleyway (or what we call the runway), which is about five hundred to six hundred feet from the second gate that we swing shut behind them. The animals are then trapped in a very small area where they are next pushed into holding pens so one cow at a time can be treated with Ivormec and a medication called "8-way," which is used to prevent infectious diseases. We herd them down an alleyway into a holding pen, then bring them in one-by-one to be doctored. I saved Butter from all that by treating her individually. While the others were treated, I kept her somewhere else so she didn't get hurt. This particular time, I had her locked in another area so she wouldn't be able to come to me while we were busy needling.

On that day, Carson brought her to me in the small area where we did the work. Well, Butter was taking Dewey's needles, taking off his cap, and, if we lay down a tag, she would have it in her mouth instantly. At one point, Dewey was filling the syringe when Butter knocked his cap down over his eyes. I think at that moment he was a little disgusted with Butter because he takes his job quite seriously. But even though she has done a lot of things to him, I've never seen him get angry with her. When I asked Carson why he had let her in, he said she needed me because she was feeling lonesome. Butter loved the excitement of the animals going through and thought it was a real adventure. She handled the commotion wonderfully, but was definitely in everyone's way and had become a real pain in the neck. Needless to say, Carson doesn't let her in anymore when we are doctoring.

In July, Carson and I went away for the weekend. Darren and Alison, our son and his wife, were taking care of the chores for us. That evening we got a call from Alison. A young calf had hung himself up in the wire behind a big gate. Alison was devastated. She had to get my brother-in-law, Darcy, to help get the calf out of the wire. The little calf was a mess: he had fought the wire so hard he had almost broken his shoulder. Since Dewey wasn't around, Darren got Carson's brother Alvin, who is a cattle rancher, to come in to help doctor the calf. They couldn't get the calf to suck a bottle because he was already used to nursing from his mother. So for two days Alvin tubed the calf so it wouldn't dehydrate.

Lessons in Elk Behaviour

Darren was beside himself. He knew the tubing couldn't go on forever. When he put down the pail of milk he was going to use to fill the bottle, the calf reached over and drank right out of the pail. Darren was ecstatic that the calf could drink by himself. Darren, Alvin, and Alison kept the little bull going until we returned and I took over. Yet another lesson had been learned from the elk herd. You may recall that as soon as they are born, the mothers take the calves to the dugout where they are taught to drink water, so the little calf knew how to drink as well as suck.

Because the calf's legs had been pushed straight back by the fence he could not move them. The first thing to be done was bring the legs ahead. I kept massaging his legs and back. When Dewey was available once again, we filled the big water tank up and put the calf in the water. While holding his head up, we forced him to use his legs in the water as a form of physiotherapy. All this time the mother was at a distance and never came to bother us. I was afraid this calf would not have a mother by the time he could get up. Butter, in the meantime, was always underfoot as we worked with the calf. She was devoted to the baby.

One evening my daughter Lana and Alison came out with me to feed the calf. Everyone was quite concerned. When I got to the calf, I lay my package of cigarettes down on the tank. As she had done since she was a baby, Butter walked over and picked them up with her teeth. When she was done playing with them she usually just dropped them on the ground. Lana, seeing her with my cigarettes, walked over to take them out of her mouth. I warned Lana to leave Butter alone, but instead she grabbed for the cigarettes a second time and Butter did the unthinkable: she raised herself onto her hind legs to attack Lana. I couldn't believe what I was seeing, but Lana had challenged her and she was going to fight back. I always carried a stick and I ran in front of Butter and hit her with it to stop her. Striking Butter almost killed me—since the day she was born I had never hit her. Down she came and around me she went, up on her hind legs again and right back at Lana. Lana turned to run but fell down with Butter on top of her. I ran and pushed Butter sideways, knocking her off balance. As I was screaming at Lana to get up, Butter went up on her hind legs again. What was I going to do? I thought that if I couldn't bring her down, I would have to throw myself on top of Lana.

She was a young mother of two babes—Sydnee (two years old) and Rachel (one year old). I was terrified! The mother has to live. I hit Butter again with the stick and down she came again—and turned on me. She was mad, grinding her teeth with her ears laid back. I ran around the water tank with Butter following me to get her away from Lana, all the while screaming at Lana to get up and run. Alison ran over to Lana to help her up and together they walked as fast as they could for the gate. I could hear Alison telling Lana not to run.

Meanwhile, Butter was still coming at me. I lifted the club to hit her again. I realized that she thought I was challenging her and was going to fight, so, with everything I had, with every bone in my body trembling, I lowered the club to the ground and reassuringly said, "Butter, Butter, it's okay." She was within two steps of me when she skidded to a halt, with the crazy look fading from her eyes.

Now, when an elk stands on its hind legs it is at least seven feet tall, give or take a few inches. Butter rose up once again on her hind legs and came down in front of me—then walked up to me trembling. I thought I was going to faint. Meanwhile, the two girls were just about to the gate. I told them to run—I knew Butter was determined to go around me after Lana. She was still mad at Lana, but not at me anymore. I started for the gate with Butter because I knew there was no way I could hold her back. The girls just barely got through the gate, with Butter right behind them. Darren, my son, had heard the commotion and screaming and met the girls at the gate holding a club. When he met Butter head on, she rose up on her hind legs again and he hit her on her nose. Down she came, running to me crying and shaking her head because he had hurt her. She stood there mewing and trembling, sucking my finger and telling me how awfully mean my kids had been to her. I stood trembling with her, trying to figure out what had brought this on.

Afterward, it was clear to me what had happened. Lana had never been in the pasture before, and while Butter was used to Alison coming in through the fence, she wasn't used to Lana. In her mind, Lana was a threat—a stranger—or a new cow in the pasture. There is definitely a pecking order in a herd and Butter thought she was guarding the calf in her own way. Lana had challenged her on her own turf and had to be put in her place. Butter didn't want to hurt Lana, but felt she had to make her give ground and run away. Believe me: Lana ran.

Of course everyone was mad at Butter. My son-in-law, Bryan, thought she should be shot. Darren told me Butter was going to kill me one of these days. Lana said she would never trust her again, and Alison warned me to be sure her kids were never inside the fence with Butter unless they were in the truck.

Again, lessons were learned: never trust a cow in the pasture, and you can never be sure of what an elk you raise will do. My scent was on both my cigarettes and the calf. Lana had made the ultimate mistake of challenging Butter, and Butter thought she was protecting my cigarettes and the calf. After this near tragedy, Lana said, "One thing is for sure, Mom, I'm glad you're my mother. I knew if someone was going to get killed, it wasn't going to be me. I knew you would fight for me or die trying. I thank you for being my mother and saving my life."

Time marched on. I still babied Butter, but noticed she was sucking my finger less often. She was now a year old, and I guess it was time for her to grow up. More and more often, she was going out to the pasture to be near the herd—still not part of them, but sticking close to them during the day and returning to the house at night.

Thunder the Bull Arrives

That fall we brought in a new bull, Thunder. He was a son of White Lightning from Easy Street Ranch. Butter was the daughter of Twister, Dwayne Veldhouse's great bull. I was hoping for a calf from Thunder and Butter because both bulls have great bloodlines. I didn't know if it would ever happen because Butter was very afraid of Thunder. When he came in she would run to her pen in the facility, which would make Thunder very angry. He was always chasing her, trying to make her go with him. The more he chased her, the more afraid she was. If I was out there with her, she would run to me with the bull on her heels. I'd tell her, "Butter you're going to get me killed." I couldn't count how many times I had to run for the gate with Butter behind me, and the bull after her.

Once by the gate, I would make my stand with Butter beside me. I would grab my club or a pail and hold it up. Thunder would then slide to a stop, I would chase him, and he would back off. I could get away with this because he wasn't in what we call "the rut" yet. This is when bulls breed the cows. Bulls are usually docile and meek with the cows during calving season. They don't dare look sideways at these grumpy, pregnant females, knowing they are busy giving birth to and looking after their babies. Bulls are expected to babysit and mind their own business, and they learn to keep their distance. They like to be fed oats away from the cows that are so ill-tempered at this time of year.

The Rut Begins and Butter Gets Hurt

Fall came along and brought the rut season. Suddenly, the docile, meek, family man—the male elk—turns into an aggressive, arrogant, overbearing bully. Everyone minds him or there's trouble. He tries to prevent the cows from coming into the facility to eat, and he works hard to keep them together in a bunch. If a few sneak away, he comes to get them and they run when they see him coming because they know they're in trouble. That year we let Thunder go to "hard horn" (we didn't cut his antlers off). His antlers were long and sharp—and he knew how to use them.

Thunder didn't care that Butter was different, she was going to go with him or he would kill or injure her. One morning I went out and there was a heart-breaking sight—poor Butter badly injured. Thunder had gored her with his antlers and she was a mess. She couldn't move her neck and was staggering sideways, bleeding from her wounds. I was sure she was going to die. Dewey came over quickly and we worked on Butter. He gave her shots and mended her wounds. She was hurt badly, and for the next week she did nothing but stand there with her head down.

After we doctored her, we went after the bull. Dewey tranquilized him and removed his antlers so he couldn't gore her again. Carson and Dewey pointed out to me that when Thunder knocked the other cows around it was okay, but when Butter got hurt—off came his antlers. I told them that it wasn't fair. He didn't gore the other cows because they knew enough to go with him. Butter didn't speak their language, and didn't know what he wanted her to do.

I kept Butter locked away from Thunder for a good two weeks. That didn't stop him. He would run at her through the fence and pace up and down determined to somehow get her. I was mad at him for hurting her, but knew I still had to expose her to him. So, after the second week, I opened the gate again. I knew he could still push her around, but he couldn't really hurt her without his antlers. People asked if I thought the bull had impregnated Butter and I would say, "Well, if he did, it was definitely against her will." I knew she had taken a terrible beating, but didn't think he'd actually managed to breed her. She had put up such a fight.

Spring Arrives—Butter Is Making a Bag

The next spring came, and, to my wonder and amazement, Butter was pregnant. I was worried for her. What if she had trouble delivering her calf? What if we had to pull it? What if she died? What if she abandoned it? What if she killed it? I had myself sick to my stomach for days on end after Carson told me Butter was "making a bag." This is a term used when a cow's udder starts to swell in the last month of pregnancy and her teats become engorged with colostrum. I followed her around, checking her milk bag every chance I got. I think Butter thought I had gone crazy. Every time she turned around, there I was behind her, checking to see how big her bag was getting. I had become her shadow; every move she made, I was there.

One day Carson said to me, "Mom, what kind of mother do you think Butter will be?" I joked, "A good one I taught her everything I know." Still, in my heart I was worried for her. She was hand-raised and I really didn't know what to expect.

On a Thursday, toward the end of June 2001, the great day arrived. Butter was in labour. But of course

anything to do with Butter was going to be different. That morning when I went out to feed the herd, Butter came back to the facility with me. I went into the house, but I could hear her calling me. I went back out, scratched her ears, petted her, and loved her the best I could. She's a big animal, but still she leaned against me for cuddling. I tried to leave, but she kept calling me, so I went and got a lawn chair so I could sit with her. This made her happy, and she went to her straw pile and lay down. It was a very hot June day and I became thirsty after an hour or so. I got up and went to the house for a pitcher of water, but before I even got my water I heard her calling for me. She had gotten up and was standing by the gate. I quickly got my water and hurried back. She went back to her straw pile and lay down again and I stayed put on the chair. When Carson came home I told him that Butter was in labour and she wouldn't let me leave her. Every time I moved she came to the gate and cried for me. I don't think he believed me because he knows elk don't want to be bothered at this time. They leave the herd to be alone while they are giving birth. But not our Butter, she wasn't going to go through this alone!

While we waited, Carson decided to hill the potatoes. The garden is right by the fence so I thought I would help him and pass some time, but the minute Butter couldn't see me she would come to the gate to call me. We finally gave up and took turns, one of us hilling a row of potatoes while the other sat where she could see us.

The day passed. Butter started moving around. She seemed uncomfortable, but nothing much was happening. At about two o'clock she lay down behind the gate with her head sticking out where she could keep an eye on us. The afternoon dragged on.

She would get up, eat a little, lie down and chew her cud. I asked Carson, "Do you think she's actually having a calf or not? She's not acting like a normal elk—thrashing around and throwing her head—and she doesn't seem to be straining." At times I believed she wasn't even in labour. Then, at about five o'clock, Butter disappeared completely behind the gate. I sensed she wanted to be alone now, so I moved away from the fence and went to the back step of the house where I sat and waited. At ten minutes to eight she came out from behind the gate and went to her pile of straw.

Foreboding—An Unusual and Ominous Birth

Twenty minutes later she stood up, and I could see her water bag coming out. After half an hour, Butter had her baby on the ground. Butter did nothing—I mean nothing normal—while having this baby. She just stayed down and pushed gently. There was no thrashing, no kicking, and no violent pushing to get the baby out. Just like that, it was there.

Dewey phoned to see how Butter was doing. I couldn't get the words out fast enough and told him how calmly she had had her baby. He was kind of quiet as he said, "Bev, you had better watch for twins." I laughed, "Yeah, right!" Heifers don't have twins unless they have been artificially inseminated (injected with the bull's semen by the rancher)—

and certainly not when they have been exposed to a bull in the normal way.

I watched Butter lick and clean her baby, eating her afterbirth. A mother elk cleans up after the birth by eating everything (the placenta, blood, and tissue). This ensures no blood is left behind for predators to smell. The instinct of the wild was showing up in Butter as she cleaned both herself and her baby. As she cleaned and licked, every once in a while I could see the baby's head looking around. When a mother elk licks her baby, you would expect its fur to get wet, but somehow the licking actually dries the fur, and within two hours Butter's baby had a new look. Her hair was dry and standing up and she was a little fur ball.

By this time it was eleven-thirty and too dark to see much. I saw the baby stand up but I couldn't see it suck. I decided I should have a bath and get to bed—I was completely exhausted. I had a good sleep, happy that my Butter and her calf were fine. Butter had had her baby now and I could finally relax.

Carson and I were up at six and couldn't wait to see what kind of mother Butter was. Carson always tries to pretend he's not as attached to her as I am, but he doesn't fool me. We were both prepared to be run out of the elk fence by a very protective mother elk as we headed for the pen. I remarked to Carson, "I guess this is going to be the end of a beautiful friendship between Butter and me, now that she has her baby." He replied, "I don't think so, Mom. After what I saw yesterday during her labour, I don't think she'll forget about you." In my heart I was not sad to let Butter go I was truly happy for her. She finally had something of her very own to love and would never be alone again.

Out to the facility we went, and what was on the ground? Not one baby, but two! Butter had given birth to twins. What a joyous moment—another miracle had happened! I was ecstatic! I ran to the house to phone the kids. Darren responded, "Well Mom, she had nobody to love and now she has two babies of her own." Lana's comment was, "Mom, God knows how hard you worked for Butter—now she's paying you back twofold."

It's a good thing we can't see into the future because there are times it would break our hearts in two. When I ran back, I was stopped in my tracks by the sad look on Carson's face as he said, "Mom, you better phone Dewey. These twins are really tiny. One can't get up and the other is quite weak." The nightmare had begun. Dewey was there in a matter of minutes. Although he doesn't like to admit it, Butter is special to both Gladys and him. The babies were premature. That was the first strike against them. I asked Dewey what had made him think she might have twins. He pointed out that because she had given birth a little too easily for a first-time mother, it could mean a premature calf or multiple births. Dewey, Carson, and Jim (Dewey's nephew) went out to Butter. I stayed back, afraid of how Butter might react.

The Struggle for Survival Begins

Butter stood up. The little heifer went to her while the men took care of the little bull on the ground. I called to Butter to take her mind off the men. She started toward me, then stopped and started to graze, keeping a close eye on the men. Butter went to the

little heifer where she stood. The tiny little thing tried to reach Butter's bag. Right at this crucial moment two strangers walked across Dewey's pasture toward ours. My back was to them and I didn't see them coming. Butter wasn't bothered by us, but when she saw the strangers she stepped back. The little calf was so weak and had had this one chance to reach the bag, but now she couldn't and she fell down.

Carson warned me to tell the people to get back but it was too late, the baby hadn't managed to reach the milk bag and was far too weak to try again.

Dewey tubed the little bull. When this was done, Butter went directly to him. She was moving from one baby to the other as we finished with them. She would go between her babies, loving them and cleaning them. Butter never threatened us, she knew us and was completely calm. She seemed to know she needed help and trusted us to do everything we could. I knew we needed good colostrum for the babies. Carson went to see Faye Grimm, who had a big Holstein cow who had just had a calf, and brought a full pail of her colostrum home. Dewey had to leave for the day and wouldn't return until that evening, so Fay tubed the babies for me at four o'clock on Friday. Alvin and Carson tubed the babies at ten that night.

Just to be on the safe side, we took the babies outside the fence away from Butter to work on them. She waited patiently at the gate for me to bring them back. Even though we were interfering, Butter wasn't heeding her natural instinct, which would be to abandon or kill her babies in a situation like this. She just waited patiently for me to bring them back, and then she would wash and clean them. When Butter lay down in front of her babies, Carson made the comment, "You know, Bev, if I didn't think she'd kick my head off, I'd hold those babies right up to her teats."

Saturday morning our babies were really in trouble and Carson said he had had enough. As Butter lay there, he picked up the calves up and lay them down at Butter's bag. I again stood back, somewhat afraid. I was not sure how she would react, but she just lay there. Carson held the babies to her nipples but they wouldn't suck. As he milked Butter to get the milk to flow into the babies mouths, he hollered at me, "Get over here! I need help!" I finally got the courage to walk over to Butter and kneel on the ground beside her and the calves. We both fought with the calves to get them to suck. Butter even lifted her leg so Carson could milk her swollen bag, but we were too late. If we had done this Friday morning when the calves were still strong we might have saved them. My fear had cost Butter her calves. I hung my head in sorrow. We worked with Butter for a long time on the ground until finally she stood up to clean her calves.

Initially the little heifer had been the strongest, but by this time she was the weakest. She had tried so hard reach Butter's bag, but was one inch too short. When Dewey came at noon to tube them again, he asked me, "Bev, do you want me to take the heifer? She is already in her own world." I asked him to leave her with Butter to die.

Losing the Fight for Life

I went into the house for half an hour but couldn't stand it, so I went back out and packed both calves into the garage. I got heating pads and blankets to keep their temperatures up. I then phoned Dewey and asked him to come back and bring his strongest stimulants to see if we could bring them around. He came and worked again, fighting for Butter's calves. Lana came and sat with the little heifer to keep it warm. Little Sydnee sat by her mother, petting the baby and asking, "Are the little Butters sick mommy?" Tears flowed from Lana's eyes as we fought to save the calves. Rachel kept walking around the calves saying "Aw, Aw!" She loved them. Lana cared for the little heifer while I fought for the little bull's life. "Please God, just let me keep one baby for Butter" was my prayer.

At six-thirty Lana had to go home. Her little girls were tired and the stress was getting to her, but as she went I knew she was crying. She had just left a piece of her heart behind for the little heifer she was never going to see again. After Lana left, I packed the little heifer out to Butter who, of course, was waiting for us at the fence. She trusted me completely with her babies and knew I would come back with them. I lay the little calf at her feet, it's breathing was very shallow. Butter cleaned her, but there was no response. When Dewey came at eight, I took the baby heifer from Butter and told Dewey to put her down. She wasn't going to live—but she also wouldn't die: it was time to let her go. We wrapped her in a blanket and Dewey left with her. I took the little bull out to Butter where she started in again, washing and cleaning. I thought my heart was going to break for my poor Butter. I left the little guy with his mother and went back to the house where I watched from the window as Butter lay down with him. She cleaned and licked and loved, but I knew Butter and I had lost the battle to keep her babies alive.

At midnight, when Dewey came to tube the little bull, he said, "Bev, it's getting pretty cold outside and the baby's temperature is falling; you had better take him inside." I packed the baby back in and wrapped him in blankets to keep him warm. I checked him at two o'clock. He looked at me with his big black eyes. I went to bed and got up at five o'clock—then hurried to the garage—only to find that the last little baby had died. I picked him up and packed him out to Butter who was standing at the fence waiting for me. I lay him at her feet where she nuzzled and talked to him. I was crying so hard I couldn't see—I thought I was going to choke on the lump in my throat. I guess I was making sobbing sounds, because Butter left her baby to come over and lick my tears, my face, my hair, and my arms, giving little elk mews. She was crying and she knew I was crying for her. As Butter licked my arm I knew she was looking for my hand. I put it up to her nose where she began to suck on my finger, something she had not done in a year and a half. Her heart seemed broken and I could feel her pain. It had all been for nothing in the end. Once again, I was all Butter had left.

I left Butter with her baby and went to dig a grave in the corner of the garden. I got the baby and took him back to the house and wrapped him in a blanket. Carson was up by that time and came to help bury him. When we were done we both went to Butter to comfort her. The three of us stood together—we had fought a battle and lost. We had all learned a lot about each other over those four days.

When Dewey phoned at eight o'clock I told him the baby was gone. He asked me if he should come over and get him and I told him Carson and I had buried him. Dewey said, "Bev, I tried to tell you. You have to prepare yourself for things like this." I replied that I knew that logically, but couldn't avoid getting attached. I just wanted Butter to have one baby, something of her own to love. Now she has nothing, no one. Carson overheard and said, "Mom, remember the day I asked you what kind of mother Butter would make? Well, your Butter made me proud. She was a better mother than I could have imagined, even if she was one for only four short days."

Next Year Butter, Next Year

When I went out to talk to Butter, she would turn her head from me as though she just wanted to be left alone. I persisted, and finally she came out of her pen and started to eat and drink. We took a walk together to the dugout, and as we walked, I told her, "Next year, Butter, next year." I wondered what lesson I was supposed to learn from this experience. I never thought Butter would live. I never thought she would get pregnant or have twins. I never thought she would let us help her. I guess what Butter has taught me is to have faith. I still think life is cruel. If God was going to put us through all this, why, in the end, would he take both calves? It seemed that, right from the start, Butter had all the cards stacked against her.

But life goes on, and next year will be here soon. I'll try to have a little more faith and maybe this fall Butter will become pregnant again. One thing I know for sure, *for the love of Butter*, I'll help her all I can.

PART 2
Coyotes, Bear Traps, and a Bull

For days now, I had been waiting for Butter at the elk gate, hoping she would come out of her pen. My hand-raised (half-human) elk had recently given birth to twins, but both of them had died. Butter was completely despondent with the loss of her calves. She was barely eating or drinking and was, I think, sick with milk fever. Her milk bag was swollen because she could not get rid of the milk. A healthy calf would have been sucking it dry. Butter had been totally ignoring me. She didn't come to me when I called her and she wouldn't even raise her head to look at me.

One Lonely, Dejected Mother Elk

I had raised and bottle-fed Butter and she was now a not-so-little two-year-old. I stood there wondering what to do. This could not go on forever. She had to eat and she had to get up and move around. So I got a pail of oats from the granary and walked out through the elk facility to her pen. She was laying there with her head stretched out on the ground. She didn't even acknowledge that I was there. I knelt down on the ground beside her to stroke her head and neck. Pity for my poor Butter welled up in my heart. Not only did the rest of the herd continue to shun her, she was now mourning the loss of her babies. I laid my head down on top of hers and cried softly for her. Butter lifted her head, sniffing my face. She has an uncanny way of sensing my feelings. She then pushed herself backwards and stood up. Once she was up she laid her head on my shoulder, sort of leaning on me.

"Come on, Butter, let's go to the dugout for a swim," I said, continuing to stroke her neck. "You're one lonely, dejected mother elk aren't you baby." Butter followed me from the pen, not even trying to stick her head in the oat pail as she normally would have done.

We went through the facility out to the pasture. She followed me, walking slowly with her head down. "Come on girl, get your chin up." But she just poked along. I walked ahead of her and reached the pond first. Usually, she would run ahead of me, bucking and hopping sideways, then literally fly into the dugout. But not today. She was completely listless. She walked into the pond, took a drink of water, turned around, and started back home, leaving me standing there by myself.

"Fine," I thought, "at least you came out of your pen for a while." I followed her back to the facility just in time to see her go back to her pen. As I watched her go, I thought how big she had become since she was born (she had been a whole 26 inches high at birth). Now, when she lifted her head above mine, she had to be at least six feet to the top of her head. At times, Butter frightens me. I am always a little leery of her—not that she had ever hurt me—but she had attacked our daughter, Lana. I knew what she was capable of doing if provoked. I try to overcome my inborn fear, especially where Butter is concerned, because I am her "mother" and she needs me, whether she knows it or not. But elk can move so fast and their hooves are so sharp that one has very little chance of getting out of harm's way if they are mad at you. As for me . . . once a coward, always a coward.

Thunder the Bull

The next morning when I looked out the kitchen window I noticed Butter was laying in her straw pile and not in her pen. "That's good," I said, talking to myself.

"What's good?" my husband, Carson, asked coming into the kitchen. I told him Butter was out of her pen, laying in her straw pile, and that Thunder had just come in and was standing beside her. Thunder is the sire of our herd. Standing over her with his head held high, broad in the chest, he looked massive, majestic. He towers over Butter a good foot and outweighs her by about four hundred pounds. Butter looked pretty small laying on the ground at his feet.

I was wondering what he was going to do, when he laid himself down on the ground beside her, not bothering her. It was like he knew she was sad. "Thunder," I exclaimed, "you can be quite a gentleman when you want to be." Thunder stayed with her a good two hours before he left to go back out to the pasture with the other cows and calves. As I watched Butter and Thunder, I wondered what kind of elk conversation they might be having. Could they both be sorrowing for their lost calves? Is it possible that the bull elk, as well as the cow, feels the loss of a calf? I guess we'll never know because they can't talk and tell us, but in my mind I think the bull elk knows when there is a loss and when the cow is upset. The question of how the bull feels would never have entered my mind a few years ago, but because of raising Butter and learning how she feels about things, I have come to believe that the bulls have affection for their babies, and for the cows they sire.

The next morning I filled my pails with oats and headed out to feed the elk. As I drove through the big gate with the old Ford I was excited to see Butter coming toward me. She likes to chase the pickup or run along beside it. She hadn't been out with me for two weeks or better. "Good girl," I said to her as she ran alongside the open window of the pickup. When

I reached the area where I dump the oats, she fell back behind the pickup and started to eat out of the pails. I hurried to the back of the truck to start dumping the oats before the rest of the herd got there. Some of the mother elk run at me the minute my back is turned, so I am very cautious when they are milling around me. Thunder also moved in to feed, so I quickly jumped into the truck. I don't take any chances with Thunder around. Mind you, at calving time, he's mild compared to the cows. They are downright dangerous.

Calving Time

I visually checked the cows' udders. That's how I tell how close they are to calving. The bigger the udder, the closer they are to delivering their calves. As I looked the herd over I could see that Number 50, a young heifer, was in labour. A heifer is a female elk that has not had a calf before. Once she has a calf she is called a cow. The cow usually gives herself away when she is ready to deliver, either by staying back or not coming to eat. She may walk the fence-line or lie close to the ground, acting sneaky, hoping you don't see her. I try to pretend, for their sake, that I don't see them. That way I help the elk feel safe. As labour pains get worse the cow thrashes around, throwing her head back and forth. Laying flat she kicks her legs, straining to get the calf out. After a lengthy time of labour the water bag, which the unborn calf lives in while it is in the womb, starts to protrude. A few minutes later the calf's front legs show up and, about an hour later, a calf is born.

I left the herd to go back to the house for a while. I would go back to check Number 50 in about an hour. As I drove away, Butter tried to follow me, but Thunder cut her off. She didn't put up a fight, just moved off away from the bull and started to graze. I giggled to myself as I drove away. I knew exactly what Butter was going to do. The minute Thunder wasn't watching she was going to sneak away from him. Nothing he was going to do was going to keep her from getting back to me. He might be bigger and stronger than her, but she was way smarter. I was right. By the time I got back to the yard and parked the truck, she was standing by the gate. Butter was getting better. Her "nobody is going to boss me around" attitude was coming through loud and clear.

The nature of the bull at calving time really puzzles me. He's not at all aggressive with the cows. He lets them push him around and he stays away from them when they are feeding. If they get there first, he actually prefers to come into the facility so he can eat away from them. At this time of the year his new antlers are growing, becoming quite massive. The newborn calves go to him, lay with him, and play in his face, bucking and bumping into him. Yet he makes no threatening moves toward them. When he is watching over them (babysitting, so to speak), he is protective of them. This is particularly evident when we are out there and I know, without a doubt, he would charge me if I got too near a calf. He acts so docile and gentle with "the ladies" that you might actually say he's henpecked. Sometimes I feel sorry for him, the way every one of his females treat him. Yet at breeding time he turns into such a horror that I feel sorry for the cows. There is no happy medium, I guess. It's tit-for-tat in the animal kingdom through the different seasons.

I was mowing the grass and as I worked I saw the bull come in and herd Butter out. No fooling around. She didn't like it, but she went with him. As soon as I was done the grass I jumped into my pickup—a different one than the one I use when I feed the elk. In this truck I have my telescope and binoculars, water, and usually I pack a lunch. Sometimes it's a very long day spotting cows having calves, as we can have up to three newborns a day.

I spot my cows from a different place than where I feed them. When I get to my spot (one that is outside the fence) I set up my scope, mounting it on the pickup's window. I am about four hundred and fifty to six hundred and fifty metres from the herd. I keep a very safe distance because the mother elk like to be alone when they are calving. The least bit of disturbance for the labouring cow puts the unborn calf in danger. The only time we move in to help a cow is if things aren't natural—such as having only one leg showing, one leg and the head, or two back feet, or if the cow has been straining for a long time and nothing is happening. The cow is then rounded up, not without difficulty, and chased into the facility. She is put into a calving box or chute where Doc pulls the calf. If Doc isn't available, Alvin, my brother-in-law, pulls the calf. If neither of these two amazing men is available, we get a vet from Peace River.

There is a lot of care that goes into these animals at calving time. I have people ask me, "How come you have to help them. When they're in the wild nobody helps them and they have their calves just fine." I say, "You're right, there is nobody to help them in the wild. They just die there on the ground with the calf partly out. Or they are eaten alive by wolves, coyotes, or bears, unable to defend themselves in any way." I don't think people realize that what we see happen to our game animals at calving time is actually the same thing that happens to the elk in the wild.

As I watched through my scope, I could see Number 50 getting up and down. "Okay girl," I muttered, "I don't want any trouble from you, just one nice, healthy calf on the ground is all I want." I turned the scope so I could see what Butter was doing. She was laying with some of the calves. "Hang in there Butter," I thought. At least the cows were allowing her to tend them. It looked like the little calves knew they had found a friend and Butter had found something to fill the void in her heart. She was doing what she had been born for: being a mother (even if it was just babysitting somebody else's babies). "Good girl," I thought and I turned my attention back to the labouring heifer. She stood up and I could see the head and feet were out and with another great heave the rest of the baby slid out, falling to the ground.

Number 50 stood there trembling, frightened at first, then turned around to sniff at the wiggling, little baby on the ground. She nuzzled, then started to lick the little one. I takes about two hours for the cow to clean herself and her baby. In that time the baby works at getting to its feet. They have quite a time learning how those long, gangly legs work. Actually, it's a beautiful experience to watch the bonding of the mother and her newborn baby.

Butter, seeing the commotion, started to wander over to the new mother. "Oh no, Butter, stay away from her!" This kind of attention doesn't go down very well with a new mother elk. I was right: the elk went at Butter tooth and hooves. Butter had probably been thinking it looked like one of her babies. All she wanted to do was have a look, but Number 50 was not having any of that. Butter knew she was in trouble. She swung away from the attacking mother and ran toward home and her pen.

Trouble Brewing

I left the spotting area and drove back home. The new mom was okay but Butter probably needed some comforting. I met her at the gate. "Come on Butter, let's go to the dugout for a swim." As I walked with her to the pond, I told her, "It's okay Butter, next year you will have one of your own and it will love only you. Just wait and see." Butter bounced back, and being the smart lady she is, wasn't going to make the same mistake twice. Now she keeps a good distance from the new moms.

Carson and I had to make a trip to Edmonton so I asked a neighbour to do chores for us. I took Laverne out to meet Butter. It was love at first sight. He liked her and she liked him. You can usually tell if Butter likes somebody, as she makes it quite clear if she accepts you (or not!). I was nervous about going because I had another cow who was near her time, but not-to-worry. Friday morning, Number 33 had her calf. Before we left, we tagged the little one, putting a green tag in her little ear—Number 14.

We were gone for four days and when we got home Laverne was worried. He said he hadn't seen Number 50's calf, nor had he seen Number 33's calf.

Carson and I headed out to see if we could find the new babies. I don't know why, but I had a bad feeling growing in me as we drove around the pasture. We rounded the cows and calves up and I immediately knew we had trouble. There was no sign of the two new calves. What could possibly have happened to them? We drove around the fence-line to check the fence. As we drove we could see where something had been digging under the wire to enter the elk pasture.

Coyotes Are Getting In!

We finally found green Number 14 laying dead, half-eaten. We couldn't find the Number 50 baby at all. We knew then we had coyotes getting under the fence and killing the newborns. The new calves are hidden in deep grass by the mothers to keep them out of harm's way. The mother somehow instills in them the need not to move an inch when there is danger. The elk then moves off to join the herd and graze, leaving the calf alone for hours until it is feeding time. She checks from time to time, always knowing exactly where the calf is. In the meantime, the coyote sneaks in through the tall grass, kills the calf, and the mother is none the wiser until she goes back to check it again. A fenced-in elk usually has nothing to worry about from the wild animals, but these crafty coyotes had dug their way in. Laverne felt really bad for us and the loss of the calves. We never did find

Number 50's calf or its carcass. We assumed the coyotes had dragged it outside the fence to eat it.

That evening I went out to the pasture to see if I could shoot a coyote. I stayed out there with the herd until about midnight but never saw a thing. I left the pasture and it was dark when I reached the gate. I drove through and got out of the pickup to lock the gate. As I was pulling the gate shut, something touched me on the back of my head. I nearly came out of my skin! Butter had followed me out on the passenger side of the truck, and I hadn't noticed her.

Butter Takes Off

So here it is—almost midnight—and I have an elk loose on the farm. There was no way I could coax her back in. With her head up in the air she took off toward our son Darren's house. I quickly drove the truck over to the granary, grabbed a pail of oats, and hurried back over to Darren and Alison's. I could see Butter up on Darren's deck examining the swimming pool. Oh Lord in Heaven, don't let her jump in the pool. I knew I would never get her out. Darren's pool is above ground, the same depth all around. She took a drink out of the pool, then jumped off the deck on the other side.

Thank God. Then she headed for Alison's flowerbeds. Some she ate, others she was pulling out of the ground with her teeth and tossing them around. I was going to be in big trouble with my girl over this one. Butter was on a roll, she could hear me calling her but she was being totally ornery. She finally realized that I had a pail of oats in my hand and came running to me. I hurried inside the fence and dumped the pail as she came running full bore toward me. She scares me when she runs at me like that. But it worked—she went through the gate and I pulled it shut. When she realized I had tricked her back inside the fence she came to where I was standing, reached out, and bit my hair (pulling it with her teeth). I said, "Butter, I would let you out but you'd chase the kids, kill the dog, want in the house, and end up in the swimming pool—or worse—end up downtown."

As I drove to the house I heaved a sigh of relief. That was all I needed . . . to get Carson up at midnight to put Butter back inside the fence. He would have throttled both me and Butter.

Coyotes—And a Terrifying Encounter

The next morning I went to the MD (municipal district) office and they issued me poison. We couldn't believe the coyotes were coming in this close to town to hunt. The MD officer said he had seen about five coyotes out in our field one morning and Carson had seen a few the morning before.

I put the poison in chunks of hamburger and

took the meat out at about nine o'clock that evening. I covered the meat lightly with dirt, planning to check it first thing the next morning. If the meat was still there in the morning I was going to take it in during the day so nothing harmless or domesticated would eat it. I didn't have to worry about that: every morning the hole had been dug out and the meat was be gone. I wasn't contending with one coyote, but a whole pack.

About four days later, when I went out to feed the herd, Butter followed me out (as usual). I was checking the holes from inside the fence that morning. First, I drove around on the inside to see if the coyotes were digging in under the fence anywhere else. On the outside of the fence the grass was high, while on the inside the elk kept it eaten-down, making it easier for me to look for new holes. I stopped to check a new hole the coyotes had dug.

As I got out of the truck I took the spade out of the back and walked over to the hole. I wasn't paying attention to anything else as I started filling the hole. Then, in horror, I suddenly realized that about five feet into the tall grass a coyote was staring at me with teeth bared. I turned to run back to the pickup, but between me and the truck was another coyote. I froze in my tracks. My heart was pounding so hard I could feel it in my throat. What could I do? I had a gun in the truck, but a lot of good it was going to do me in the predicament I was in. I was scared silly, my legs wouldn't move (even if I willed them to) so I did what I do best when I am scared. I screamed bloody murder! I was screaming Butter's name. If anything or anybody was going to help me right then, it was going to have to be her.

She was grazing ninety metres away from me. Up came her head. Butter had never heard a sound like that come out of me. Without the slightest hesitation she started to run toward me. I don't think Butter had seen the coyotes at that point. She was just coming to help me and she covered that ground fast. In the meantime, when I screamed the coyote by the pickup made its move—running at me. I started to run for Butter and, in that moment, Butter saw the coyotes. I swerved around Butter, not knowing if she was going to run right over top of me in her rage. She was mad! Teeth grinding, ears back, and barking like a dog. When there is danger the cow elk make that call deep in their throats. She went around me knowing exactly where I was, missing me by centremetres, and started to pound on the coyote with her hooves. The coyote rolled away from her. Around Butter came—making a turn—coming right back down on top of it. The second coyote ran for the hole under the fence. Butter, seeing the second coyote, left the one she was pounding on. It was rolling and scrambling to get way from her. Butter went after the one trying to get to the opening. Up on her hind legs she went, coming down hard on the coyote.

All this time I had been trying to dodge Butter and the coyotes, working my way to the safety of the truck. I knew I had to reach it before the rest of the herd got to us or I would get trampled. Butter, up on her back legs, had the second coyote by the opening and was stamping with her front hooves. There was so much fur flying and snarling going on behind me I didn't know who was getting the worst of it. I hoped it wasn't my Butter. As I reached the truck, taking a look back, I saw the coyote crawl through the opening. It looked like Butter had broken it's shoulder. By this time the rest of the herd had reached us and I had grabbed the gun. Meanwhile, Butter had turned back on the first coyote who was trying to make it to the hole. It didn't make it. The herd had it. Down under their hooves he went. Once he came through, only to be pawed down again. I sat in the pickup shaking. There was no sense in me trying to shoot at the one dragging itself across the field. With all those elk stomping around me and the truck, it would be my luck to shoot something I didn't want to hit.

With the rest of the herd finishing the job, Butter came over to the truck, tossing and throwing her head around. She was blowing through her nostrils, quivering all over, trying to get air back into her lungs. I reached through the window of the truck as she stuck her head inside. I wrapped my arms around her neck, crying and kissing her on her bloodied nose where the coyotes had got a few of their own licks in. "Thank you Butter, thank you God. What would have happened to me if you wouldn't have been with me?" Would those coyotes really have attacked me . . . or would they have turned and run?

I guess we'll never know. All I know is that Butter knew I needed her. I'm pretty sure, with her uncanny sense about me, she knows I'm a coward through and through. And when she heard that god-forsaken scream come out of me, it must have scared her half to death. She must have known that I was in some sort of big trouble—trouble that I couldn't get myself out of. She responded to my cry for help the same way that a mother elk does with her calf.

There wasn't much left of the coyote when the cows were done with their stomping and trampling. I left the herd, with Butter following behind me, and headed for home. Me to my bedroom to pull myself together, and Butter to her pen to nurse her nose. We'd had enough for one day.

Just One Bad Thing After Another

I was pretty downhearted that afternoon as I sat down to do my books. It hadn't been a good spring at all. Butter had lost her twins. And, we'd had to pull Number 74's calf, because it was born dead. I didn't know how Doc Dewey was ever going to get it out of the cow because its little head was twisted. But Doc did it and saved the cow. However, Number 23, a beautiful three-year-old cow, had aborted in March which is way too early to have a calf. By the time we realized she was in trouble and pulled the calf, we were too late. We ended up losing the cow too. I couldn't believe this had happened because I watch them so closely.

My hands were full. A friend of mine, Jimmy, was building a new fence so I was calving out his elk along with mine. His first little bull that was born was so spunky I named him after Jimmy. He was my favourite. Jimmy liked his food, and this cute little baby always tried to act like a big elk. Right from the start he was eating grass and nibbling at oats. I just loved him. One day I saw him walking really slowly, his head down. I phoned Jimmy and told him to come in so we could check the calf. Doc Dewey also came over to check him out. He had an infection of some sort in his intestines, and within a few days we lost the little guy. I was heartbroken. He was three weeks old and there was no known reason for this to have happened.

Another cow of Jimmy's also had trouble calving—one leg and the head were out, but the other leg was not showing. We tried to run her into the facility, but in no way could we get her in. She would just run back to the herd where the rest of the cows and calves were running like crazy. I said, "That's enough, Jim, we'll phone the vet to bring a dart gun and put her to sleep before we kill another calf running them all like this." The vet had to come from Peace River, an hour away from Manning, because Dr. Dewey was gone for the week. It was about four in the afternoon by this time, with the cow still in its predicament, and about nine o'clock in the evening before the vet got here because he was so busy. I was sick with worry that we would lose the cow before we could pull the calf. By that time, the cow had been in labour, with the calf partly out, for over six hours. To be honest with you, I was in agony waiting for that vet. I was so relieved when he drove into the yard.

The veterinarian from Peace River put the cow to sleep and we pulled the calf. The vet and the men didn't even look at the calf, presuming it was dead. And it looked dead—its tongue all swollen, with no movement or sign of life. I thought, what the hell, I'm so tired of things dying on me. The vet had a pail of water sitting on the ground by the cow that he used to wash his hands. I reached into the bucket and quickly washed the calf's nose off and put my mouth over his nostrils. I started blowing air into its lungs. I quit and started pumping on its heart, then back to blowing in its nose. I felt a flutter of life as one of the little legs gave a faint move. Back to his heart I went, thumping on his little chest—back to blowing into his little nose—he took a gulp of air, then another. The vet went running for some sort of injections, gave the shots, and the calf started to respond. Jimmy couldn't believe it. He said he never would have thought to breathe into its nostrils. He had wondered what the heck I was doing. Carson said, "Oh, I knew what she was up to—she'll try anything." The vet patted me on the back and said, "I'm sorry, I thought there was no life left in him." Jimmy asked me what made me try after so much time had passed. I said, "Jim, when you want something to live bad enough you'll try anything. What did I have to lose?"

When the mother elk came out of her sleep she wouldn't accept the calf. So Jimmy and his family took it home to bottle-feed. Jimmy helped some other neighbours out at the same time and between them they had three babies to bottle-feed that summer. A lot of work for everyone, but Jimmy has four daughters and everyone pitched in to help. I think Corrina, his wife, was the happiest when they weaned the calves and turned them out to the herd.

Later, as I sat there filling out my registration forms to send the government to receive birth certificates for calves, I was feeling pretty woeful about everything, including the coyote ordeal and a forty thousand-dollar bank payment looming in the fall. Things were looking pretty bleak. Then I heard Butter calling me. She wanted some attention. I walked out to her and wrapped my arms around her neck and gave her a big hug. Whether Butter knows it or not, she comforts me too, especially at times like this. She doesn't know she costs me money and gets me into trouble, she just loves me. As far as she's concerned, that's what mothers are for. As I stood there scratching her ears, I said, "Well, Butter, our baby calves are drawing every coyote in the country to our pasture. I think we need more help than just the poison."

Butter and the Bear Trap

Later, I phoned a friend who traps and he brought me a Cona bear trap. Mark set the trap in the hole for me, on the outside of the fence on the ground, where the elk would not get into it. Butter was there watching us set the trap. I didn't think much about it. I went back to the house when Mark left, but the more I thought about Butter watching us set that trap, the more it bothered me. So I jumped in the truck and drove back out to the trap.

I could see Butter over in the corner where the trap was set. "What the heck is she doing," I thought, as I drove up. "I have to get her away from there!" I skidded to a stop, my heart in my mouth.

Oh my God. "You little dummy!" I screamed. She had put her nose right in the trap and she was caught. She was down on her knees fighting to get away from it, her nose and mouth were bleeding and she was making it worse as she fought. The ground around her was all pawed up with her thrashing around, but the trap was holding her tight. She was a mess—trembling, snorting blood though her nose, and struggling to breathe. Thank God it hadn't cut off her air altogether.

There was no time to get Carson. I would have to try to pull the trap apart myself. Crying, I reached through the wire of the fence calling her name, hoping she wouldn't pull farther away from me. Her eyes were full of fear and pain. I reached for the trap, pulling it apart with every ounce of strength I had. Her nose came free. But because she was pulling her head away from me with such force, she went straight over backwards and fell to the ground. I couldn't go to her because I was on the opposite side of the fence. Butter just lay there sucking air into her body and then—up she got, running as hard as she could for home and her pen—not even bothering to thank me.

I knew she hadn't been in the trap for more than half an hour, but that was long enough. She was lucky and I was lucky; she could have died in there. I jumped into the pickup and raced back home as quickly as I could, grabbed some wound medicine, cloths, and water, and went out to her pen where she was hiding. And did she have a story to tell me! I cleaned her nose and chin where the trap had cut into her and sprayed the areas on her body where she had beat herself up. She put her head on my shoulder and rested. "Oh, my poor Butter, you are always getting in trouble." But it was my fault . . . again. Butter is just too curious; she had seen us over there setting the trap and simply had to check it out. Another lesson learned. Don't do things Butter can see, because where I am she will be, and, in her mind, if I'm there it can't be bad.

Counting Our Blessings

Time marches on and every day brings its blessings and curses. I often think of all the people who have pitched in to help me with the elk and of the wonderful neighbours we have. I think of how many times my brothers-in-law, Alvin and Curtis, have dropped everything to come and run a cow in and help pull a calf, the loads of oats and bales that they have supplied to feed our animals, and the help they have given when we needle the herd. Another brother-in-law, Gary, came for a visit from Fort St John, British Columbia. He is the manager of the lumber mill over there. Gary ended up helping pull Number 74's calf in the spring. A man who has never been around animals got right in there with the vet and helped turn the calf's head around to get the calf out of the cow. Then there's Carson, who would rather be trucking, our son Darren, who has his own job and responsibilities, and Jimmy Gordey, who comes for miles at a drop of the hat. Dwayne Veldhouse, who hauls my animals from one end of the country to the other (and who is my best friend in the whole world), raises the greatest bulls you have ever seen. Last but not least: Darrel Hunter, who raises my bulls for me. Then, of course, there's Doc Dewey. Well, there is absolutely no way that man could live without me. I have become the hangnail in his life. He would miss me and my chocolate cake if I disappeared. And he can't chew me off.

One day, Carson was disgusted with me and the elk about something and I didn't have a good come-back to his remark, so I said, "I just don't understand: I have about ten men in my life, counting the vets, and you guys can't seem to be able to look after one little woman." He gave a snort and told me, in no uncertain terms, that I was harder to look after than fifty women put together. I am blessed with the men in my life. No matter how much trouble I seem to get myself into, my men seem to get me out of it. So, now you understand that Butter isn't the only one who gets herself into trouble around here.

Elk Ranching—A Hands-On Learning Experience

About mid-August we brought the herd in to give them their needles for the winter, and at the same time we separated Jimmy's animals from ours so he could take them home. I was relieved to have that job behind us because there is always the fear that a calf will get hurt or trampled. Two days after we had done the needling, Carson came in and said, "Mom, I see an elk down in the corner standing by herself. You had better go out and check on her." I drove out the to pasture and, lo-and-behold, Number 296 had put a calf on the ground. It was a little late in the season, but the baby would be big enough by the time the winter months came. I phoned Jimmy about the new calf and told him I had had a late calf, and that he should watch his Number 27 cow who hadn't had a calf yet. And, maybe she was making a bag.

About a week later I got a call from a very upset Jimmy. He had built a fence around his oat field and had gotten his brother to come over to swath it. His brother, while swathing, had run over a new baby calf and he was devastated. He told Jim, "If you would have just mentioned to me there might be a calf I would at least have been watching." Jimmy told his brother that I had called and told him to watch for one, but that he hadn't given it another thought. It didn't cross his mind again, even when his brother went out there.

The swather had cut the baby's leg off and the vet told Jim to put the little one down as there wasn't much that could be done for it at that point. I felt sorry for Jimmy. He didn't cry. But I knew he felt like it. These things are hard on us elk ranchers. We try so hard, and sometimes the harder we try the worse things get. There's much to learn about our wild animals. One of the hardest things to get through our heads is that a cow can sneak away and have a baby, hide it, and rejoin the herd. And *nothing* about her appearance is any different than the last time you checked the herd. Other people can tell you what to watch for, but overall it's a hands-on learning experience. We all walk around saying, "If I had just," done this, or "If I had just," done that but it makes no difference. Because the next year or next problem is totally different. But it's a given, and it will make your hair stand on end.

The Great Dugout Caper

Fall arrives, and the rut with it, (this is when the bull breeds the cows). The bull becomes very aggressive and much more dangerous. Breeding involves impregnating the cows so that calves will be born in the spring. At this time, the bull keeps the cows bunched and doesn't allow them out of his little circle. The cows aren't allowed to come to the facility without the bull's approval. If a few slip away, he's after them immediately. Needless to say, Butter and I keep as far away from him as we can. Butter for her reasons and me for mine. One afternoon, Butter was standing at the gate. She looked bored, so I said, "Come on Butter, let's go to the dugout." She followed me out. The bull and the rest of the herd were

out of sight. I thought Butter and I were safe; the herd doesn't come to the pond till the evening, so we had lots of time to go there and get back.

Butter ran on ahead, leaving me behind. I laughed at her as she showed off for me, giving little bucks and prancing sideways at imaginary things in the grass. The pond has high banks at its sides so, as you go down to the water, you can't see over the top. Butter was up to her tummy in the water when I reached the bottom of the dugout. She took off swimming, the water cooling her hot body on this a warm, fall day.

I stood watching her—as she turned to swim back to me she gave a warning bark—the call elk make when their calves are in danger. I thought, "What the heck's the matter with her?" It didn't take long for me to realize what the problem was. There, on the top of the bank, was Thunder, the bull. My hair stood on end. "My Lord help me," was my thought, as he came charging down over the bank, trumpeting, straight toward me. I didn't know where to go, so I ran straight into the dugout toward Butter. As the heaviness of the water slowed me, I ducked under, screaming as I went. The last thing I saw as the water closed over my head was Butter turning to swim away from the bull.

As I went under the water, I could hear and see the bull's body next to mine as he crashed into the water missing me by inches. He was so close to me I could have reached out and touched him. My lungs were screaming for air. I had to come up to suck air into my lungs. As I surfaced I was praying I wouldn't come up beside him. But when my head broke the water, he was right in front of me. I took a quick gulp of air and dove under the water again, going to my left, away from him. I surfaced the second time. I was pretty sure I wasn't going to come out of this one alive.

I had to do something. Face it—I'm not a good swimmer and the water was fairly chilly (even for a warm day). So I frantically headed for shore. When my feet touched the bottom I took a chance and looked around. Butter had gained the bank on the right side and the bull was swimming toward her, ignoring me. I hauled myself up on the bank, scrambling as hard as I could. Once out, I ran as hard as I could for the fence, praying that Butter wouldn't run to me for protection. If she did, there was no way I'd be able to climb that eight-foot fence. I'd have to jump right back into the dugout.

I turned to look over my shoulder at Butter. She stopped to look at me, then at the bull, then started to run for the facility (and her pen) with the bull after her. I stopped running and tried to get a grip on myself. I was sobbing from sheer terror. I had nearly gotten myself trampled (and drowned!). I stayed by the fence. I couldn't go to the facility with the bull in there, and I couldn't go to the big gate in case the bull caught me out in the open. I had to wait by the fence until the bull came out, either with Butter in tow or by himself. And God help me if Butter decided to come back to me.

Butter must have made it to her pen inside the facility because after about half-an-hour the bull came out by himself. I made not the slightest movement. The mood he was in would be my death warrant if he came back for me. I'm really not sure I could have climbed that fence. But he sauntered out, not bothering with me at all. Once he was out of sight, I lost no time dashing for the facility. Butter came to meet me, licking my wet hair. "Sure, Butter," I said, "you're just as big a coward as I am when Thunder chases you. But at least, for a change, you ran the right way and didn't get me killed. I almost drowned! But I guess I can forgive you for that—but you can forget about me going to the pond with you anymore; you're a big girl now, and you're on your own!"

When I got to the house, Carson was just getting home. Seeing the mess I was in he wanted to know "What the hell!" had happened to me. He wasn't that impressed with Butter and me. He put his foot down. No more walks with Butter! And absolutely no more pond swimming! Mind you, after he calmed down he started to laugh, "That must have been quite a sight, you and two elk . . . swimming in the pond." Yes, Dad . . . it was quite a sight . . ." I muttered under my breath as I walked away from him. If he only knew how traumatized I was. Jeeze, I was actually happy to be alive. In fact, it felt real good catching what-for from him.

You're a Real Piece of Work, Butter

The following weekend, my sister Louise and

brother-in-law Gary came to visit. We had gone shopping for most of the day and it was getting quite late when we returned home. Butter was waiting at the gate when we drove into the yard. When she saw me she started running back and forth. Louise asked why she was doing that. I said, "I haven't spent any time with her today and she misses me. Louise said, "You know something . . . she scares me." She acts half human." I laughed and said, "I know the feeling." But she's spent most of her life with a human family. It's hard to explain to people how she acts. "Believe me. She thinks she owns me!" We walked over to Butter and Louise asked, "Is she cranky? She's acting like she's mad at me for taking up your time. It's almost like she's thinking, 'You took my mommy away from me . . . and I don't like you.'"

I turned and looked at Butter through someone else's eyes and couldn't believe what I saw. She was actually acting snotty. Her head was up and she was looking at Louise out of the corner of her eye, pretending she wasn't there. All the while, she was trying to get closer to me. When she finally got real close she started pulling my hair with her teeth. She does this when she's not pleased with me. When she's happy she licks my arms, when she's disappointed she pulls my hair with her teeth, and when something really bad happens to her she sucks my finger (just like she did as a baby).

As I stood there listening to Louise, with Butter pulling my hair, I realized something about my precious Butter. She really was a spoiled brat. Finally, having my undivided attention (and totally ignoring Louise) she turned her back to Louise and pushed me in front of her. It worked. I was away from Louise and she had me all to herself. I went and got some oats for her. She likes to lie down and eat them. I said to her, "You're a real piece of work, aren't you Butter. Everyone who gets to know you points out something else weird about you. But I guess I'm stuck with you. Besides, what else could I do with you? I could never eat you, and you know darn well it's true. But you better be good, because that's what some people would like to do with you, especially Bryan."

Caught With Her Hoof in the Cookie Jar

In the fall, Carson and Darren bought a load of oats to dump in the feeder. It's open on the sides so the elk can just stick their heads in to eat. We fill the feeder in the fall so that when the snow is deep they can eat on their own without me packing pails to them. Then, at the beginning of February, they are taken off their oats so that they slim down for calving. This is important. If a cow is too fat, she will have trouble birthing. I try to ration the cows so I can still give them a few oats to keep them calm when I am working around them.

The men backed the truck up to the grain auger hopper basket that catches the oats and carries them up into the feeder granary. Of course, Butter had to be there to help them. As the oats were coming out of the little chute on the back of the truck, Butter caught them in her mouth. Then she tried to eat out of the basket where the auger was whipping the oats up the sieves into the granary. Darren, being afraid Butter would get her nose sliced off, shooed her away. She jumped sideways and pretended she was afraid of him. Finally, he had to give her a smack to make her mind him. She left the auger and went over to where the motor with all the belts were turning. Well . . . she puts her nose up to the belts and they slap her good a few times, stinging her severely. She closes her eyes when they hit her and jumps sideways: "Gee, this isn't much fun." She stands there a few minutes and licks her nose with her tongue, a little blood showing where the belt had taken some skin off. "Well, I wonder what these are?" as she sidles back up to where the motor ropes are hanging. The rope is wound around a pulley then pulled, which starts the auger motor. So, ever looking for excitement, Butter pulls on the rope with her teeth until she gets the rope off—then she takes off with it in her teeth. Darren runs after her to get it away from her. She can't lose that rope or there'll be big trouble with the men running after her. Butter is very happy. Someone is finally playing with her. Darren grabs the rope from her and gives her another smack. It took her a few seconds to register that Darren was mad at her. Then, head down, acting like he had beaten her to death, she stood pouting, looking at him as if to say, "Boy, are you a party pooper or what. A girl can't have any fun around here with you around."

Darren went back to the truck to lift the grain box up. Carson said later that the minute his back was turned she scampered around to the back of the grain truck to catch the grain in her mouth. She was having a gay old time when all of a sudden she realized Darren had come back and caught her. She acted like a little kid with her hand in the cookie jar—she jumped straight backwards and sat on her haunches. Darren started to laugh, so, just like a little kid (sensing he wasn't really mad at her), she ran over to him to get him to scratch her ears. The men were glad when they were done dumping the grain. Butter had made a fairly easy task into a real challenge.

Butter ate so many oats that day that whenever I went over to pet her all she could do was burp and hiccup in my face. I told her that her manners were not acceptable, but that didn't even phase her. She carried on burping and letting gas go (too many cookies, I guess).

A Serious Accident—A "Thank God" Outcome

It's not just the elk that keep us in a turmoil; even small animals can cause serious problems. One day, our little granddaughter, two-year-old Rachel, picked up a stray tomcat. It attacked her, pulling the skin on her forehead completely off. The weight of the big tom—not being able to retract its claws when Rachel dropped him—peeled that baby skin right back. We had to send her to Edmonton by emergency air ambulance. Once there, it took two hours of surgery and over a hundred stitches to repair Rachel's forehead. I stayed up all night talking back and forth with my sister Sandra who had driven into the city from Dayton Valley to be with Lana and the little one. Rachel's dad, Bryan, arrived later by road because he wasn't able to go on the plane with her and her mom.

By morning light I was completely distraught over the whole ordeal. Trying to be strong for everybody else took its toll on my own emotional state. I knew my granddaughter was going to be okay, but I couldn't get rid of the lump in my throat. It seemed as though it was lodged there. So I grabbed a coat and headed out to Butter who was waiting at the gate as usual. As I reached out to pet her she sensed how upset I was. She maneuvered her big body as close to me as she could get. As she sniffed my face she gave a pitiful little mew. A dam broke inside of me when she made that mournful sound. I wept into her neck; she nibbled on my hair, and I told her all about Rachel: "Butter, we're sure having a bad year with our babies, aren't we?" As I stood there, Butter kept giving me her little mew. Butter, in her own way, was trying to tell me it was going to be okay. And it has been. Rachel is fine, aside from her scar, which will heal and fade with time. Amazingly, she still loves cats and is not the least bit afraid of them.

The Fall Rut—Butter and the Bull

Poor Butter's body was going through a change again. She didn't know what was happening to her. She wanted to be with the bull but she was afraid of him because he was so aggressive. So Butter would come to the gate and call for me. When I didn't go out to her right away, she would run up and down the fence-line. Back and forth she went, and I'd feel so sorry for her that I'd go out to calm her. It wouldn't be long before she would go looking for the bull or he would decide to come for her. My only hope was that he wouldn't get too rough with her when he tried to make her go with him. Last year he gored her (and almost killed her) when he tried to breed her. Now it was the end of September and she still hadn't been bred.

One morning in early October 2001, Butter wasn't at the fence waiting for me. I ran out to her pen. She wasn't there. But what a mess everything was in. In the area where we do our doctoring, shelves were torn down, equipment strewn all over, and gates were pushed open and the wooden ones smashed. Thunder had gotten into the holding area to get to her. I was frightened. Maybe she was down somewhere, really hurt. I ran to the pickup truck and took off out to the pasture to find her. If Butter had gone into heat and accepted the bull everything would be okay, but if he was forcing her before she was ready it could be a disaster.

I got to the pasture and spotted Butter right away, grazing with Thunder, and looking no worse for the wear. For weeks he had been chasing her, but he must have known she wasn't ready to mate. Somehow, last night, he knew it was her time and went to get her. I am amazed by that bull's intuition. Thunder has thirty females to breed in the pasture but he never forgets about Butter, up here by the house, and still senses when she goes into heat.

Of course, when Butter saw the pickup she wanted to come to me. Things got a bit rough then; there was no way Thunder would let her leave. I put the truck in drive and drove away from her. There was no sense making it harder on her. I could hear her crying for me. I got a lump in my throat for my four-legged baby. I knew I had to leave her till the bull was ready to let her go. If we wanted a calf in the spring, this was the only way it was going to happen. If Butter was ever going to have somebody in her life besides me, we had to get a calf on the ground for her.

Elk Grieve Too

Time flew by. The middle of November came and Dwayne Veldhouse and Darrel Hunter were coming to move my spring bulls down to Darrel's bull farm. We were running the herd down the alley to the holding pen when Number 4 turned and ran back. Because she was frightened, she turned too sharply and ran into the corner post. She hit it so hard she spun herself around and down she went. I knew she was hurt but she got up, so I was hoping it wasn't too bad. When we were done separating I went to check on her. She had gone to the far end of the pasture and was laying down. I took her some oats and water but when I got close to her she got up, so I left her. The next day Carson walked her up closer to the house so it would be easier for her to get food and water. The morning after that she was gone again, and I went to find her. This time she couldn't get up. I phoned the vet clinic, and Gladys, Doc Dewey's assistant, came over. We gave Number 4 shots for the swelling and pain. We worked on her for the next three days, trying to make her get up, but we lost her on the fourth morning. When I went out to check, my beautiful animal had died.

Butter, who is a pain when we are trying to do something, had been locked in the facility while Gladys and I worked on the hurt cow. I finally let her out of there and she followed me out to Number 4's body. As I stood there sadly, not believing that this beautiful elk had really died on me, Butter walked over to the cow on the ground and lay down beside her. Butter stretched her neck over and gently licked the cow's face as if she was saying goodbye. Elk, like elephants, grieve when they lose their calves, and the cows are sad for days when we take their calves away from them. Butter was showing me that they also feel the loss of their friends (not that the other cows ever welcomed Butter, but Butter always wanted to be friends with them). Now was her chance to mourn and to show that she cared about the other elk not that they had ever welcomed her (and to show that she was, after all, half elk). It was touching, and I cried as I watched Butter. I left them lying there together to go get Carson. He would have to bring a loader over because we had to remove the cow and bury her.

As I drove away I was thinking about Butter, wondering why I would even want her to have a calf and what problems it would bring for Carson and me. Butter has never been run down the alley with the herd into the holding pens. When she has a calf, what will happen to it? It will be wild. How will we get it to go down the alley without Butter? Will Butter take it down herself, or will we have to separate them? And if it's a little bull, it will be taken away from her when fall arrives. We simply can't keep the bull calves. They have to go to a bull farm because there's only room for one bull on our elk farm, and that's the sire. If it's a little heifer we can keep it, but chances of it being a heifer are slim. I just shook my head as I drove to get Carson. Maybe he is right when he asks me how much more of this I can stand. I looked back at Butter as I drove away. And in that moment I knew in my heart how much I do have to stand—*for the love of Butter.*

A Face-to-Face Encounter with Thunder

I went out some time later to do my chores. It was one of those days when even the sunshine didn't help. As I headed for the troughs to dump the pails, the bull lunged at me through the fence. So I went further down, on the outside of the fence, to throw oats to him. This move would force him to go around the dividing fence to eat. I knew the rest of the herd would follow him around, giving me enough time to open the gate and put the oats in the troughs. But as I walked in, the gate swung shut behind me. I saw Butter running toward me as I lugged the pail. She only comes to me when the other cows are away from her because they delight in chasing her back to her pen when she's alone. As I dumped the pail, Thunder made a sudden turn from the cows and came charging back toward me. He covered the ground so fast that I had no time to think what I should do or where I should go. I dropped the pail and jumped over the first trough, then cleared the second one (and I always thought I couldn't jump). Amazing what you can do when you fear for your life. The bull hit Butter, spinning her around in front of me. She regained her balance and took off for her pen as I reached the gate.

There was no use trying to follow her to safety because the bull would outrun me. He could easily kill me. I grabbed at the gate and brought it toward me, knowing I had no time to get through it. Backing up, I used the gate as a shield. As the bull hit it, he knocked me right back into the fence. He was in a hell of a mood, running at the gate over and over, but I kept swinging it out at him. I was trapped. Even though he did back off, he wouldn't back off far enough so I could duck around the gate and get out. Butter knew I was in trouble. She would run half-way to me and the bull would charge at her. This scene was repeated several times. Thunder kept his eye on me, coming back to me each time after he charged Butter. Finally, after about fifteen minutes of this performance, he walked over to some oats laying on the ground. But he did not take his eyes off

Butter—or me. I wondered how long he was going to keep me there. At the same time, the rest of the cows were coming back in to eat and the gate was still open. As they came close, I kept swinging it at them to keep them from getting out.

Eventually the elk ate their fill and started to leave the feeding area, but not Thunder. He was having too much fun running Butter off and keeping me behind the gate, pinned to the fence. As the cold started seeping through me (it was twenty-six below zero, with a wind)—my fingers already frozen from handling the pails—I was beginning to wonder if anyone would ever come looking for me before I froze to death. Butter kept trying to help, but she was only making things worse. As long as she was there the bull was going to make sure she didn't come to me. Finally, after about an hour, Butter decided to head over to her feed stack. As soon as she was out of the bull's line of vision, he followed her far enough away from me that I could made my move—ducking around the gate and quickly pulling it shut behind me just in time.

He saw me move and charged, causing me to lose my grip on the gate when he hit it. I reached for my club, which was standing by the post, and swung it at him, hitting him right on the side of his jaw. Boy, did that feel good! I told him, "I feed you every day, look after you, and you treat me like this! Now how do you like being treated like a pile of manure!?" One thing for sure: Thunder made up my mind for me that morning. I am going to have to bring in another bull. Thunder is getting too dangerous and aggressive for me to handle. He won't back off, and I can't take the chance of him hurting one of the men when we have to handle him. The elk have to be somewhat afraid of us so they will back off. And clearly, Thunder has no fear.

That gave me cause to think why, when you have bottle-fed elk, they can be very dangerous. And that's because bottle-fed elk have no fear of humans. If you anger or provoke them, you don't have a chance in hell against them when they come after you. So I worried a little with Butter, as she munched her oats and burped in my face: "Wouldn't that be something—to go through all this with you—then lose you."

Another Calving Season—Bread, Butter, and Toast

Another spring has arrived and we are about a month away from calving time, patiently waiting to see what lays in store for Butter. Right now she runs away from other cows. But when she gets over to where I feed her (by the horses), she becomes really tough and grits her teeth, running at the fence that protects the horses and showing them she is the toughest, meanest elk on the farm.

During the third week of May 2002, Carson's favourite cow, Eve (Number 238), had the first baby of the season. This was a real surprise to me because she still had last year's calf sucking her. I wasn't even watching her till I noticed her run her yearling calf off. I went inside the fence to look around and, sure as heck, I found a newborn calf. Her last year's calf and I became buddies. The little one would come right up to me to eat out of the pail. She was lost because her mommy wouldn't let her come near her. Once the new baby got bigger, Eve let the first calf come back to her.

A few days later, two cows, Number 78 and Number 76, were ready to have their calves. In the distance, through my scope, I saw two feet coming out of both cows. Everything is going well. At around two in the afternoon, Number 78's calf was born. I waited for another hour. When Number 76 stood up I could see she hadn't made any headway, so I went in for a closer look. The calf was coming backward. At four o'clock I went to get Carson and Darren and we ran her into the calving box.

I phoned Doc Dewey, who was in High Level at the time, to explain the situation to him. The cow had been jumping around and had jammed the baby's leg that was sticking out. I reached inside but I didn't know what to feel for. Doc said I had to find a bum or a head. I panicked when I couldn't find either. We lost Doc on the cellphone, so Darren phoned Alvin. Alvin said, "It's okay, you have to reach way in—the calf's back legs are as long as your arm. . . ." And I had only reached in up to my wrist! By that time, Alvin was sure the calf was backward. Alvin tells Darren, and Darren tells me, to pull with the cow when she has her contractions. "Pull down and pull hard!" I started to, but knew it was going to be a hard pull. Carson stepped up behind me, put his hands on top of mine, and we pulled together. I heard Darren answer "No, . . ." then he said, ". . . pull fast, you guys, or the cord will wrap around the baby's neck and choke it!" We pulled. And out it came. The calf couldn't get up because it had a broken back leg. That meant there was no way the cow would take it or look after it.

Carson brought the calf out to me and it looked at me with those big black eyes and bleeped at me as if to say, "Give me a chance, I have nobody to love me. Help me." I put the gun down and picked the baby up and packed him to the garage. I said, "Look, little fellow, I have Butter. I can't call you Bread cause I don't think you're going to make it. But it's this way: if you try I will try—and if you give up—you're toast. I phoned Doc Dewey to come and put a cast on the broken leg. When Doc arrived I said, "I'm probably wasting both your time and mine. He'll probably be toast by tomorrow." Doc took the baby

over to the clinic and later brought him back to me. On his cast, Doc had written *Toast*. He had no ear tag yet, but at least he had a name.

Doc let me fight all day to get the baby to suck. It makes me so mad. Every other elk farmer sticks a bottle in the calf's mouth and it sucks. Not for me. So the battle began. At nine o'clock, Doc tubed the little guy and said, "You're not getting enough fluid into him and he's becoming dehydrated. His temperature is starting to fall." So it began. I was running between trying to feed Toast and checking the cows in the herd for problems. Number 73 put a baby on the ground that evening with no problem. Doc tubed Toast again at midnight, but before he did I tried again to get the calf to suck. After that I set the alarm for four in the morning and fell into bed. The clock was right by my head but I didn't wake up until six o'clock. I hit the floor running, frantic with worry about the calf. Carson heard me cursing, "The damned alarm didn't work!" He picked it up and said, "Mom, it's wound right down. You must have slept right through it. You're pretty tired." I replied, "You have to be joking!" I have never slept through an alarm or a telephone ringing in my entire life! (I must be getting old.) Needless to say, the calf was no worse for wear. I forced the bottle on him again, and again no luck.

The next day, Carson and I went out to tag Number 73's calf and Number 78's calf, which we couldn't catch because the cows would either get the calves up or run us off. It was fairly early when we went out, and right along the fence line there was a calf lying with no tag.

We have started tagging inside the truck now because it makes it a lot less dangerous than me trying to keep an angry cow and the rest of the herd off of Carson. We got to the calf with me driving the pickup alongside the calf until Carson's door was right by the calf. He jumps out, scoops the calf up, and we roll up the windows and put the tags in their little ears. As Carson went to put the calf back out it squawked, and around the corner came Number 78! She looked just like a barrel horse coming around the corner of that fence. Carson had just enough time to jump back in the truck. Number 78 was so mad she ran at the truck. I had to laugh at her. She was mad at herself for making a mistake like that—leaving her baby unattended. But it didn't end up very funny. As I started to drive away she attacked the truck again, and I ran over her foot. She limped around for a few days but ended up no worse than before. One mean mama, that one.

Carson took Toast outside, where he learned to stand on his three legs with the cast on the fourth (he more or less dragged it). He looked like a little drunk staggering around, but he was up and trying so hard, still not sucking, but I kept trying. Number 1 and Number 71 had their babies later that day. Between calving and caring for Toast, I found myself on the run. I was so stiff and sore from getting up and down with Toast I could hardly stand. But I knew I'd get broken in eventually. Later, when I said my prayers that night, I asked the Lord to teach Toast how to suck.

By this time, Butter was making a bag and Doc was not happy with me. He had told me I let Butter get way too fat, that she might have to have her calf pulled. "No more oats for her till her baby is born!" Carson had been warning me about over-feeding Butter; but it's hard—she's always standing by the fence calling me.

We went out and tagged Number 1's calf and

Number 71's calf the morning after they were born. It went well, even with that crazy Number 78 still chasing the pickup. She was limping but was still madder at us than a wet hornet. I wished I could give her a calm-down pill of some sort. When we got back to the house, I mixed some formula for Toast, who was still refusing to suck. Doc came at nine to tube her. Toast was so pitiful to watch, dragging that splint around, but was beginning to use her leg more and more. I told Doc that morning that I felt like the most useless person in the world (never being able to get those calves to suck). "Oh no," he says, "you're the best neighbour I've got." Then he looks around and says, "Oh . . . I guess you're the only neighbour I've got!" I have to laugh at him—he knows when I'm on the verge of crying (or having a nervous breakdown!). That evening, at the six o'clock feeding, little Toast still wouldn't suck. I was tired. I was mad. I was sore all over. What was I going to do?

Sydney, Rachel, and Lana stopped in to see the calf and everybody prayed that it would suck, and that its little leg would heal. A little later, Brittany and Morgan came over and told me they also had said prayers for the little elk. Brittany had been coming over in the mornings before school to try to help me feed the baby. My sweet grandchildren were totally devoted to this struggling calf, each of them knowing that if it didn't suck soon it was going to die. I came to the house holding the bottle full of milk. What to do?

Then I thought of my dad. He never fed a calf with a bottle; he always straddled the little one, put his finger in the calf's mouth and pushed its head into a pail of milk. It was either drink or drown. Suddenly inspired, I grabbed a bowl from the cupboard, poured milk into it, and went back to Toast. I shoved his little butt end into the fence so he couldn't get away from me and pushed his head into the bowl. We spilled some, but I was sure most of it went down his throat. I came to the house and showed Carson. He said, "You know, Mom, I was thinking of trying that too." At nine o'clock I made a double batch of milk, put it in an ice cream bucket, and went to feed the little one. The whole family came to watch to see if he'd eat out of the pail. I pushed Toast's head into the pail. He snorted around a bit with my finger in his mouth, then he sucked that milk up so fast he looked like a vacuum cleaner. Everyone was clapping their hands and laughing. I looked at my grandchildren and thought, "Yes, the Lord says if you have the faith of a child and ask, it will be done." God had answered the children's prayers. Toast was eating on his own. Now I turned to my next problem. Butter.

Since I had taken her oats away, she had been standing at the gate crying pitifully for me. I said to Carson, "Just listen to that. How is anyone going to get any sleep around here?" Darren phoned over and said, "For God's sake, Mom, go give her a handful of oats so she shuts up for the night." She continued to cry. I would try to run out and pet her but that would make her happy for about five minutes, and then she'd start calling again. I whispered in her ear, "This is what they call 'tough love.' No more oats till your baby comes! Be good now. Give me break. You'll live through this."

Brittany rode over on her bike the next morning to help me feed Toast. In the evening, we took the baby elk back into the garage so nothing would harm him during the night. My old housecoat still hung on a hook in the garage over the calf's head. Toast was always up, bunting and biting on my robe, in a hurry to eat. Because my scent was on the robe, he would try to make it feed her. He was so darn cute. Everybody was in love with him. When we were done feeding him, Carson would take him outside and, just like a regular baby elk, he'd find his favourite spot in the deep grass to hide and to sleep till the next feeding.

I hurried out to feed the herd and check whether any of the cows were in labour. Then I went home and had a bath, tidied the house, grabbed the grocery list, and headed up town to get the mail and run a few errands. Back home to put the groceries away, and then out to check the herd. Came home

again, had a quick lunch, then mixed Toast's special elk formula. Lana and the little girls had come to help grandma with the feeding. We walked out to Toast. I helped him up and instantly knew there was something wrong with him. He hung his little head. I asked Lana to phone Doc Dewey. While she was phoning, I packed Toast in to the house with a dread that came from deep down. I had seen this before and knew what was coming. Because the necessary formula creates the wrong colostrum, elk babies are susceptible to infection in their intestines. We worked on Toast, keeping him warm with heating pads and blankets, turning him so pneumonia wouldn't set in, tubing him with Pedolite, but to no avail. Toast just got weaker and weaker. That night I lay down beside him, putting my arms around him and holding his little body next to mine. I told him if he wasn't going to stay with me that it was nice to have gotten to love him for a few days and that I would keep his picture in my heart and memory. Toast died around one in the morning. At three, I was still awake trying to think of a way to tell my grandchildren. How do you explain death to children that young? How do you explain why the baby didn't live? Pretty hard to do when I felt so desperate myself. Their mothers will have to help me.

Grandpa Carson dug a grave for Toast and we buried him that evening. Our little grandchildren—Brittany, Morgan, Sydnee, and Rachel—each had a flower to plant on Toast's grave. As we went to bury Toast, the sky opened up and it just poured tears for a few moments. We planted the flowers and took some pictures by Toast's grave. As we stood there, God put the most beautiful rainbow in the clouds for Toast—his way of telling us that he was with us. Each little one had a prayer to say, but the one that sticks in my mind is, "Please God, give Toast a new mommy in heaven with real good elk milk. Grandma can't seem to get the milk right." (That one was Sydnee's.) A friend of ours (who also raises elk) phoned the next day, saying he had found a note from Brittany in his daughter's knapsack, asking them to come to the elk funeral. He was inquiring when the funeral was, because Brittany hadn't mentioned a time. I felt bad and said we had buried the little elk the night before. Wade said he wished he had found the note the day before, as they sure would have come.

A Fine Balance—You Lose One, You Save One

Jimmy Gordey phoned one night quite upset. His favourite cow, Harriet—who I think is special as well, because of her calm attitude—had had a set of twins, but Jimmy had lost both of them during a terrible delivery. When he had gone out to check her there were four feet sticking out. One twin was coming the right way, and the other twin was coming backward, lying on top of the first one. They had to push the calves back in and untangle them, then pull them from her. They both died in the process. Harriet made it through the ordeal and is healing nicely. Some of our experiences during calving time are very sad, but the consolation for this encounter was that if Harriet had been in the wild she would have died that way, with no one to help her. This way she's alive with somebody to care for her and nurse her back to health, so she can give another baby next year.

Three more of our elk had calves that same day. Things went well and all three were tagged, but my mind was on Butter, who was bleating steadily for oats. She wasn't hungry, because there was lots of green grass. I was hoping that she would lose ten pounds before the birth of her calf.

A few days later, Number 3 was having her calf but was clearly in trouble. Carson and I had tried to run her in but she ran right in among the baby calves and we were afraid she would run over one. It was getting dark, so I phoned Dewey, who is on his way home from High Level. I asked him if she would be alive come morning if we couldn't get her in. He said yes, but it would probably be a dry birth, which is harder on the cow. I told him to bring the dart gun in the morning because we would have to dart her. We knew there was no way we could catch her to bring her in. Darren joined us in the morning and, as Doc arrived, I lost it and started to cry. He put his arm around me and said, "We'll get her one way or another. If we can't get her this way, we will shoot her and end her misery. We tried to get close to Number 3, but she just kept running into the herd. Darren went back to the house and got the quad. Then all hell broke loose. He separated her from the herd and kept her running as we all moved into position so she couldn't get by any of us. Of course, when she was in that predicament, and not able to get around us, she headed straight for the facility with Darren right behind her. They got her into the calving box and Doc went straight to work.

A Sense of Humour Gets Us Through

Darren started to rib Doc as he worked, saying, "Gee, Dewey, Mom didn't have to use a chain to pull her calf. Do you want her to help you?" Just then, out came the dead calf. Doc reached in again and he started to pull another calf out. Darren looked at me and said, "Mom! You didn't check for a second calf. What if she still has one inside her?" Dewey shook his head and said, "Darren, it'll be dead by now." But when the second calf came out, she was still alive. Doc draped it over the fence, so the mucus could run out of its lungs and nose, then turned back to the cow, reaching in again. Darren asked Doc what he was doing, and Doc Dewey said, "looking for a third one." I thought Darren was going to faint but asked Dewey, "How come we're having so many backward calves?" Dewey replied, "I guess the cows are pointed the wrong way when the bull is breeding them." Darren said, "Oh . . ." Then it dawned on him what Doc had said and we all had a good laugh over that one. Of course, Doc has such a dry sense of humour, and if he can get a joke in, or cause a little trouble, he loves nothing better. When he comes over and has to tranquilize an animal, he always offers a dose to Carson first because Carson's always damn mad over "Bevy's elk." When the elk are having problems they are mine. But they're his elk when everything is going fine.

We left the calf with the mother for twelve hours. She didn't hurt it, but she wouldn't touch it. Carson finally went out and brought it to me and all I can say is that I tried as hard as I could to save it. Doc did his best for me. He came in and said, "She looks like Butter, acts like Butter, and I guess we should call this one Margarine." So Margarine became her name. This little one, as ironic as it seems, started to suck . . . to my joy! I was beginning to think that maybe, just maybe, I could save this one. But on the second day I knew I was going to lose her too. I wish someone would invent elk colostrum just for me.

I slept with the little one in my arms, as I knew it was failing fast. As I slept I had a dream. In my dream the baby elk was sucking her bottle; the sun was shining brightly on her so that her little body was gleaming in the sunshine. The only thing different was that it wasn't me feeding her. It was Jesus. I woke with a start as the little one took her final gasps of air and departed from this world. Something stirred inside my spirit, and I said, "Lord, when I leave this world . . . please, let me look after the elk in the next."

Baby Butter Arrives—A True Celebration

Butter went into labour on 6 June 2002, at the same time as Number 60. I closed the gates on Butter so she couldn't wander out, just in case she needed help delivering. Then I went to check on Number 60, who had her calf and was busy cleaning it. I left her and went back to Butter again. At three o'clock the baby's feet came out. At six o'clock, after three hours of hard labour, I was convinced Butter was in trouble.

Doc wasn't home because he was up north with the travelling clinic. I phoned my brother-in-law Alvin and explained the situation. He didn't think we could leave her much longer, so he said he would have supper and then come in. Fear set in. Total panic that is. Dewey had told me she was too fat and that we would probably have to pull the calf. His words had become reality. I phoned the golf course for Carson to come home, checked Butter again, and then went to water some plants. I couldn't bear to watch her struggle. At times she would lie on her back with her legs straight in the air, trying to push the calf out.

The men arrived and we all walked out to Butter. I started to cry, saying, "I can't do this one, guys." Alvin simply said, "You have to. She knows you. Maybe you can keep her calm." We walked up behind her and there on the ground was her calf. We backed off to see if she would bond with the calf, but she just stayed down ignoring it and straining. Alvin said, "Sometimes when they have had a hard delivery they can't quit straining . . . or else she's got another one in there."

We watched her for about an hour, then Alvin

said he was going to check for a twin. As he went back out to her, she stood up and Alvin reached inside her to check. Only Butter (supposedly a wild animal), who was not even locked in a calf box, would have allowed somebody to do that to her. There was no second baby, but Butter was still ignoring the calf. By this time the calf was standing up. Butter lay down again, then got up, and out came the afterbirth. As soon as it hit the ground, Butter turned to her baby and started cleaning it.

The baby wobbled here and there, then found Butter's teats and started to nurse. Best of all, she was a little heifer. Believe me, I shed tears I was so relieved that everything had turned out so well. I said to Alvin, "Would you come next time when I call wolf," and he answered, "Oh yeah . . . this is the first time you've called that it turned out we didn't have to pull one." I suppose the guys thought I had overreacted because it was Butter, but, on the other hand, they all know she's pretty special to me. We named the calf Baby Butter. It seemed the only thing to do.

A Million Dollar Memory

Everything was going fine. Then, when the baby was six days old, I went out to feed the herd in the morning and saw Butter down by the dugout and said to myself, "Good girl, you're teaching your baby to drink and swim, just like I taught you." Then I left to tend the main herd. When I was done there, I went to help my sister finish cleaning our mother's garage out because we were moving her to the lodge.

We were just finishing up when my cell rang. It was Carson, who told me our neighbour had phoned to say there was an elk out. I said, "It must be Butter's calf." But Carson had asked Robert if it was a little one and Robert replied, "No, it's a big one with two yellow tags in her ears." Butter was out—and so was her calf.

Apparently, the frost had heaved the post up by the dugout and there was a two-foot opening under it. It looked like the baby had rolled under it and couldn't get back in. So what else could Butter do but wiggle out through a two-foot opening to get to her calf. She left a fair amount of hide behind, but managed to get that big body out of there to protect her baby. Carson was quite concerned because everyone has dogs over that way. He thought the calf was history. Thank God for our special neighbours. Everyone put their dogs in to help us out.

Then we went for Butter and her calf. We caught the baby up by the fence and put a cover over her face. Carson packed it to the pickup with Butter right on his butt, but she didn't attack him. He laid the calf in the back of the truck and hopped in with it, and we took off for home. As we came down the road past Dewey's clinic, Bev Warren, who works at the clinic, watched us go by. Her first reaction was shock, then disbelief, and then she started to laugh. Bev said later that her first thought was, "Oh my God, one of Bev's elk is out." Her second was, "Lord . . . it's Butter following the truck." Her third thought was, "Why don't I have a movie camera with me. I could make a million dollars on this one."

Butter didn't like the pavement very much, but she didn't leave the back of the pickup where Carson had hold of her baby. Once Darren, who was driving, turned in toward our place, Butter seemed to recognize her home road and ran up alongside the driver's window. Darren thought she was going to try to attack him through the window. I said, "No, Darren, she's just making sure you turn in at the right driveway." Darren drove right in to the back gate where Butter lives. I jumped out and opened the gate. Carson hopped off the truck with the babe in his arms and laid the calf on the ground. Butter ran in right behind him, sniffing her calf. The baby jumped up, ran to the granary, and lay down.

Butter watched her go, walked over to the water tank, and took a long drink of water. She then turned to me standing there, washed me all over, sucked my fingers, looked at her baby once more, and walked off to the facility. As she disappeared around the corner out of sight, she stopped and looked at me as if to say, "Mom, you look after her.

I have a headache, and I've had all of that kid I can take." I whispered after her, "Now you know some of the headaches you caused me."

The next night the herd came in to check things out. And Butter's calf went out with the herd. Butter was somewhat upset. We drove out with the green pickup to see if we could find her calf for her. All Butter had in her head was that we had her kid in the back of our pickup. She kept looking in there instead of calling for her calf, so we chased the herd back her way. She finally found her calf and peace reigned.

Wild and Wonderful—Letting Butter Go

Baby Butter is wild. So far, it stays away from us, hiding when we are around. When I went out to feed them this morning, Butter and her calf weren't around. They were out at the herd. I've been watching Butter's reactions. She doesn't like it—but the calf prefers her own kind—Baby Butter follows the herd and the other calves. As I watched, I thought to myself, "Yep, that's the way life is: the mother goes where the kid goes, whether she likes it or not. Butter has no choice: she will go wild too, for the love of her calf. She can stand there and ring her hooves all she wants, but where her calf goes Butter will go, and she will grow."

Butter's natural love for her calf and her instinct

to protect her little one casts all fear out of her. I guess God's been teaching me to have faith in Him, and in Butter. And now He's teaching me to let Butter go. But one thing's for sure—*for the love of Butter*—I'll do everything I can to help her and Baby Butter.

PART 3
One Human—One Elk—Friends for Life

The sun was shining but the air was cold. Carson, our son Darren, big Jim Gordy, and Rex Coupland (an inspector with the Canadian Food Inspection Agency) walked out to the facility.

Testing for Disease—A Nerve-Racking Time

Rex's job is to test our animals for diseases like bruccellosis (tuberculosis). This test is done every three years. First the elk's neck is shaved, then blood is drawn. We wait three days and repeat the procedure with each elk to see if we have a reactor (this would mean one of our elk has the disease). This can be a very nerve-racking time. We have to worry about cows running into posts or a calf breaking a leg or getting crushed in the crowd. Last year an elk (Number 4) hurt herself so badly she died from her injuries. Little did I know this year was going to be our hardest yet.

My own elk, Butter, doesn't come down the runway and into the facility with the rest of the herd. When I want her in, I always take her oats and she follows me to her pen. She is so curious that if we brought her in with the others she would want to be right where we are so she wouldn't miss a thing—sticking her nose in as usual.

As we walked into the pasture where the elk were milling around, Butter followed me. I ignored her as we started the push with the herd. They tried to run back the way they had come to break through us, but we ran toward them. They turned around and headed to the corner we were chasing them into, toward the runway.

It's Every Man for Himself

It was rutting season, however, and Thunder, our miserable bull, was feeling cocky. He stopped just short of the runway to run at Carson. I skidded to a stop, looking for the closest gate or fence to climb if he got by Carson, or killed him (I'm not kidding, this elk is mean).

It felt like abandoning the ship when it is going down—but at that moment it was every man for himself. It's not as though I hadn't run out on Carson before when the herd was coming down on us. As I turned to run, Darren and Jim ran past me hollering and yelling. By this time, the herd had turned in the runway and were thundering back down the runway toward us. With my heart pounding in my chest, I watched the men turn Thunder. He had decided to go back to protect his cows from us. As Thunder ran down the runway, the cows skidded to a stop to turn with him and go the way we wanted them to.

I heaved a big sigh of relief as I watched the men swing the gate closed behind them. Once the elk are trapped in the runway, we have to run them down into the working part of the facility. While this procedure is quite dangerous, at least they can't run loose and escape.

Butter Keeps an Eye on Things

In all of the excitement I had forgotten about Butter and Baby Butter. I turned to see where Butter was. Of course she had stayed out of the way of the herd and was on our side of the fence. She had seen all this performance before and seemed to love the excitement, even though it scares the rest of the herd silly. I was relieved to see her. Baby Butter, however, had not been babied or brought up like her mother. She was as wild as the rest of the calves. Baby Butter ran with the herd, leaving her mother behind.

There is an opening for us to go through the facility once the herd is out of the runway. We entered the work area, put up boards to block the opening behind us, and went to work. Carson and Darren worked the gates, pushing the animals forward, forcing them into holding pens. Rex, Jim, and I shaved, needled, and took animal inventory. Once Carson and Darren had the holding pens full, they returned to help us. One by one the animals were tested. It's a very slow process and some cows put up a fight, slowing us down even more. Once the elk are doctored, the gate is opened and the animals run to freedom. Who should suddenly appear in the work area but Butter.

I hear Carson's voice, "What the heck? How did she get in here?" No problem for Butter. Where we had come in and blocked the entry, she squeezed through to get to us. Butter was worried and started calling out and mewing, as if asking us where her calf was, and why she wasn't coming out. I walked over to pet her through the gate that divided us. I told her, "Don't worry Butter, your baby will soon be out." We returned to our work, ignoring Butter. As each elk was set free, it would try to get away as fast as it could. Each time a cow or calf bolted to freedom it would scare Butter and almost run her over. She would take off and return only when the fleeing animal was completely free.

Out of the corner of my eye I noticed her leaving, then coming back, taking for granted her patience as she waited for her baby to come through. When the fifteenth animal came through, Butter decided she had had enough of being ignored. She had asked nicely for her baby's return and we hadn't helped her out, so she took matters into her own hooves. She flicked the gate handle up with her nose and walked into the work area to look for her baby. Carson, Rex, and Jim were busy with a cow but the look on Darren's face was priceless. We ran over to her, pushing her back through the gate. Once we had her back far enough, Darren pushed the gate closed and hooked the chain.

Darren said, "Mom! Butter opened that gate with no problem at all! She can get out any time she wants if we don't keep the chain hooked on. I didn't know she could do that." I shook my head at him and asked him, "Why do you think I tell you, 'always make sure to hook the chain when you do chores'?" Butter is way too smart for her own good when it comes to fences and gates. She could set the whole herd free one day. I told Darren not to feel bad and that I just take for granted that he knows all the little things Butter can do. I giggled to myself.

I wasn't going to tell him about the time I was cleaning the water tank and forgot to hook the chain. I went to fill the pail with oats, but wasn't concerned because I knew I would be right back. I drove over to the granary behind the big shop. When the pails were almost full I heard something coming. I thought my daughter-in-law's horses had got out. But looking up I realized it was Butter. She had flicked the gate handle up and was as free as the breeze. Jumping into the pickup, I drove around the shop and back into the yard by the house. In order to dump the pails into the fence, I have to back up the pickup. No problem, Butter just followed the pickup around until I stopped, and then walked to the back of the truck and proceeded to eat the oats. As I walked around the truck and began to remove the oats, one pail at a time, she got greedy. Butter began to realize that I had more oats inside the fence than she had in the truck, and that didn't sit well. That would never do! Back inside the fence she came, and I closed the gate.

The Departure of Thunder the Bull

We had no further mishaps with the testing of the herd that day. Three days later, we ran them in again. Rex inspected every one. We had no reactors, giving us "clean herd" status. This allows us to move our elk to other farms or sell them over the next three years, when they will be tested again. We loaded Thunder up, to move him to an elk farm in Wildwood, Alberta, before we set the herd free. I was pulling him for this year because there was no market for elk at that time and it was a big-time low for selling them. To save time and pasture, and to avoid more problems down the road, we weren't breeding cows this fall . There wouldn't be any calf crop in 2003. Thunder and I parted company easily. While I respected him, he had put the run on me more times than I could count. He didn't care if it was the rut or not—he is an aggressive bull and you could never turn your back on him.

The bull farm is a five-hour trip, so the great Thunder was one tired elk by the time he was safely delivered to his new home. I didn't think I would feel bad leaving him there, but as we drove away, tears came. I felt a little better when, on one last look back, I saw him beating up a tree.

Butter—Friend and Confidante

One evening of 27 September 2002, the phone call I had been waiting over a year for finally came. The news was that my Butter book would definitely be published—but not for at least another year. The disappointment was awful because I thought the book would be released by Christmas. I had talked to all my friends and family and I felt a little embarrassed. Sometimes I felt like people did not believe my book would ever be published. I made a vow not to mention it again until the published book was in my hands.

Crying, I grabbed my coat and headed out to the facility to find Butter. I was so upset. I was thinking about the past summer and about another little calf named Fudge that I had pail-fed for two months. About a week before, little Fudge got in Thunder's way. He had stomped Fudge, breaking his back, and I had to put the poor little elk to sleep. Crying, I called again for Butter, hoping she would come to me when I needed her so badly. The air was cold as

I stood in the black of night feeling sorry for myself. I cried for my sister, Louise, my mother, my kids, and my grandchildren who all believed in me. They were all going to be so disappointed for me. How could I ever tell them?

Butter wasn't coming. Since her calf had been born, I came second, so I cried about that too. I called a few more times as the cold fall night chilled my old bones. As I turned to go, a nose touched my cheek, making me jump right out of my skin. How does she creep up on me like that? I hadn't heard her coming, but then I had been crying out loud. I decided I was acting like a baby and cried about that too.

I wrapped my arms around Butter's neck and cried, "You have to wait another whole year before you become a star." I wanted the whole world to know about Butter, whose only aim in life is to belong somewhere. She didn't fit into her world and she didn't fit into mine. I, at the very least, hoped that she would fit into a book. Butter nuzzled my cheek, licking my tears. She put her head down so I could scratch her ears and down her neck to her back. Butter doesn't know my words, but she sure understands when I'm hurt or sad. She lifted her head back up and I laid my head on her neck. Butter is so big now! This wonderful, understanding elk let me rest there in the dark until my sobs subsided. Eventually, I couldn't think of anything else to cry over.

I could sense Butter was starting to get restless. She gave a jerk and a mew at the same time. Her calf (Number 15) answered her back in the pitch black of the night. The calf had finally followed Butter in. Baby Butter wanted her mother. I patted Butter one more time and pushed her away: "Go to your baby, I'll be okay." Turning away, I headed back to the house and Carson. Surely I would get some sympathy there. Life is full of sadness and disappointment and I was no stranger to them.

I shivered as I headed for the house. It was getting colder and colder by the day. The dugout was frozen over, so the herd had to be watered by hand. Tomorrow I would fill the water tank. Once it snowed I would quit watering the herd because they would eat snow through the winter. I pulled out the heater that kept the water from freezing in the tank and put it on my to-do list for the next day.

Baby Butter's Close Call

I was standing in the living room watching the herd from the window when I noted a cow acting peculiar. I got a pair of binoculars to see better. The cow (from this distance, I couldn't tell which one) was running frantically up and down along the dugout. I hurried out to the pickup and went off to the elk as quickly as I could. I had to stop to open and close the gate. I drove like a mad woman toward the dugout, sure that a calf had broken through the ice. It had—and that calf was in the water fighting for its life. It was Baby Butter. Of course, the cow that had gone completely crazy was Butter. I didn't even stop the truck, I knew I had to get Carson's help. I wheeled around and drove like a bat-out-of-hell to get him. Carson was in the shop as I tore into the yard with the pickup. He knew something was wrong and ran out to meet me. I blurted out what was happening and he turned and ran to get a couple of ropes, and then ran to the truck and dived in.

We headed back to the dugout, leaping out of the pickup before it stopped. Carson was cussing every cuss word he knew as he tied the rope around his waist and headed onto the ice. The calf was fighting as hard as it could to stay above the water. Carson had the second rope in his hands and was throwing it to the calf, trying to get it around her neck. At the same time, he was hollering to me not to let go of the rope tied around his waist in case the ice broke beneath him. As Carson threw the rope one more time, it settled neatly around the calves neck. I heaved a big sigh. It was a beautiful throw. But the relief was short lived. The ice broke beneath Carson and he disappeared under the water. I frantically pulled on the rope and up came Carson. I pulled as hard as I could. Carson was hollering at me to pull harder. He wasn't letting go of the rope he had around the calf. They were coming in slowly, with the ice breaking around them. Finally, Carson got footing under the water and started to haul himself and the calf in. Butter was still running around like a lunatic. She ran in front of me, tearing the rope from my hands. This caused Carson to fall back into the water again. As I grabbed for the rope and started to pull again he came up swearing. You would not believe what he said he was going to do to me and the elk! I'm sure he was swearing under the water. As Carson and the calf regained the bank, the frightened and exhausted calf started to go crazy, fighting the rope. Of course, Carson couldn't let go. Finally she gave up, lying there as if we had choked her. I hurried over and slipped the rope off her. She was alive, but so shaken up and tired she was not going to move. Butter hurried over to her, and I to Carson.

Butter—Almost Dog Meat!

I had to get Carson to the house as quickly as possible before he froze to death. Carson got into the truck and I drove. I said, "Well, if that wasn't one for the books!" He misunderstood and thought I meant it was a good story for my book. Well! Carson's gentle blue eyes turned red and I swear the steam from his clothes was actually coming from his ears! I have seen Carson get mad before, but this was a total fit.

He hollered at me at the top of his lungs: "Do you

really believe anyone on God's green earth would believe the trouble you and these animals get me into? Do you think for one minute anybody else in the world would put up with this stuff, let alone believe it?" As I got out to open the gate, he was still hollering, and when I returned the tirade continued. Thank God I had to get out once more to close the gate. I had almost gone deaf but didn't dare tell him that.

When we got back to the house I tried to comfort him with humour, telling him, "Just tell everyone that the woman you married is causing you so much trouble you're going crazy. The people who know us would certainly believe you, especially your brothers and your mom and dad. They always believe the best of you and the worst of me, so you shouldn't have any problem convincing them." He told me to be quiet.

Carson did not talk to me for the next three days, so I talked to Butter. I told her, "When you girls were worth twenty thousand dollars each, and a spring heifer was going for twelve thousand, everybody couldn't get close enough to an elk. Now that you aren't worth anything, nobody wants you around. 'When you laugh, the world laughs with you. When you cry, you cry alone.'" At this moment I was alone. "You're lucky to have me for your mistress, Butter, because if Carson had his way, you would all be dog meat. This time he is really holding a grudge against us. I hope he gets over it pretty soon or I will have to tell him he is acting silly. I will tell him I really like his blue eyes and miss them. Maybe that will sweeten him up."

An Unexpected Visitor

The last week of October was cold but sunny. Morgan's sixth birthday was the first of November and fell on a Friday. Saturday morning I got up and did my chores. I filled the water tank, Carson helped me with the oats, and we left to watch Morgan's first hockey game. I got home first and noticed the garage door was open. I thought how strange that was as I walked into the house. I wondered who could have left the doors open and was getting a little cross that my family couldn't even close a door behind them. Winter is not a good time to go leaving doors ajar. As I entered the living room, I got a horrible feeling that I wished I lived somewhere else. There—laying in the middle of the living room watching TV—was Butter. She glanced up at me from her position on the floor, as if to acknowledge me, then resumed watching TV. I have antique lamps on top of both the TV and the fireplace mantelpiece. I walked slowly around Butter to avoid scaring her and rescued my lamps and a few pictures. I moved everything I could out of harm's way. I opened the patio doors to the living room and then sat in the big chair watching

Butter, waiting for Carson to come home. I was trying to figure out how she had got out.

I suddenly realized, with horror, that the rest of the herd might be out. I went out the back door, and indeed the gate had been opened. Thank God her calf was with the rest of the herd or she would have been in the house with Butter. I went in and shut the other gates so the herd could not get out, but left the first gate open so we could get Butter back in. I realized I now had a bigger problem, because Butter had flipped both the gate latch and the chain. I took a pail of oats to Butter to lure her out. She looked at me through the open door and yawned, stretching her legs out into a more comfortable position and resumed watching the television.

I could hardly wait for Carson to come home, yet I was worried about it too. I had just recovered from the last tongue lashing from his near drowning in the dugout. I thought he was going to kill me this time! I thought I might try crying when he got home so he would feel sorry for me and not get too crazy, but I could not squeeze a tear out. And my time was up. Carson was home. I met him at the front of the house and told him we had a problem. He folded his arms and waited. Then he started to wave his arms to get me to say it. I took the plunge. "Butter is in the house watching TV," I blurted out. "No way" he said, and walked past me into the house. I know there is a wrong time to laugh, but for the life of me I could not cry. Therefore, I burst into giggles. I could not believe myself. Didn't I have any sense of self-preservation! And I couldn't stop. I told Carson she was deeply involved in the program and he better not touch the remote if he didn't want a real fight on his hands. If you understood what a remote control freak Carson is, and how much he hates to share his TV with anyone let alone an elk, you would know how hilarious this was to me. Carson then started to laugh because I couldn't get a grip on myself. He threw his arms around my shoulders and asked me what we were going to do. I told him I didn't know; maybe we would just have to wait until she got bored and wanted to go back outside on her own. We did not want her to get upset and wreck the house. We also had no idea which door she would choose when she did decide to leave. Carson walked up to her and kicked her lightly on the rump, telling her to get up. Up she stood. She looked a lot bigger indoors! I held my breath, wondering what she would do.

Carson walked out the garden doors and Butter followed him out. He jumped off the step, which is about three feet high, and she jumped down right next to him. Carson picked up the oat pail and she followed him to the gate just as if it was an everyday occurrence. We had lived through yet another family ordeal, and had managed, again, to stay married. Butter, in all her worldly wisdom, didn't know how close she was at times to giving Carson and me heart failure.

Butter Frees the Whole Herd

It was 17 December and everything had been going so well for me. The elk were happy, Carson was out working, the kids and my grandchildren were fine, and I was getting ready for Christmas. The phone rang. It was Lana. She had fallen down the stairs with her baby Ashley, who was five months old. Bryan had taken her to the hospital. Her ankle was badly sprained and she had done a number on her tailbone. Ashley was strapped in her car seat and was thrown from her mother's arms, landing upside down. Lana was horrified as she tipped the baby chair right-side-up. But to everyone's relief and amusement, Ashley, unhurt, had apparently enjoyed flying through the air. She grinned as if she was inviting Lana to go through the whole escapade again.

The next morning, I went to help Lana out. I was over there about three hours when the phone rang. It was a neighbour of ours letting me know he had just driven by the farm and there was an elk out. He knew it was not wild because he had seen the tag in its ear. I could not believe it was one of my elk, but as I drove into our yard I realized my whole herd of elk was in our neighbour's field. My daughter-in-law, Alison, was waiting by the yard for me. We quickly made a plan before trying to get them in. We couldn't afford to make any mistakes or we could end up with elk uptown or on the highway. To make things worse, there were no men around to help us out. They were all in Peace River and we didn't have time to try to round up anyone else. We had to figure it out on our own. We loaded oats in pails and I started across the field to the elk. Alison ran and opened the big gate by her house.

As we were working, my niece Tanya arrived. She had been driving by when she noticed the elk in the field. She placed her car in the road so they could not get around her and, if they headed that way, she could steer them off. In the meantime, Alison started around the herd to push them toward me and the oats. I stopped the truck, got out, and dumped some oats, calling for the herd. When they saw me they headed for the pickup. As they got close to the end gate, I drove away from them hoping they would follow me. But, instead, they turned and ran straight over to my neighbour's farm. I got close to them again, stopping to dump oats right on the tailgate. Eve, Carson's pet and his favourite, came first. Then her calf. I drove very slowly so they could get a mouthful of food. Thank goodness these animals love oats so much because if you tried to push them they would bolt, and God help us then! As the last elk came through the gate behind me, Alison slammed the gate closed. This time I was able to cry without even trying!

Once again, Butter was the culprit—flipping the chain and lifting the latch, then biting through the tarp strap. Now the gate is under lock and key. The really ironic thing is that Butter and her calf never even came out. She had taken her baby out to the pasture where she was safe. Maybe Butter could stroll out, but she knows her baby is not allowed out of the fence. She acted guilty, and when she saw me come in with the rest of the herd she took off running. Now if you know Butter and her love for oats as well as I do, you would have to agree with me that here was one guilty elk!

What scared me the most was that some elk farmers had threatened to set their herds free. The lack of pasture, combined with zero market for the animals, had some farmers' backs against the wall. If I hadn't been able to retrieve my herd, people might have thought I'd let my own herd out. Who would believe that Butter was responsible? It didn't seem to faze Carson, who just went around telling everyone that he lets the elk out and I put them back in. He thinks it's a great joke!

Butter Is Becoming an Elk

One Sunday morning in January, Carson went out to do chores. That morning he was also trying to find an

easier lock for me to use on the gate. He opened the gate to dump some oats and Butter walked right in front of him. He reached up to push her back to go around her, but she raised up on her hind legs, flailing him with her hooves. When she came down she was gritting her teeth (a bad sign). Carson reached out and slapped her hard on the side of her head. She went back up, balancing on her legs for a good thirty seconds, this time keeping her front hooves to herself. She wasn't going to risk using her front hooves and getting slapped again. When she stood on the ground again, she backed away from Carson so he could dump her some oats. As usual, he was not amused.

The winters have been quite hard on Butter since I have not been feeding her separately from the herd. She has been forced to push and fight for oats with the other elk. I always dump the first of the oats by her, but if the other cows push her away it becomes her problem. She needs to learn to be tough, to stand up to the other cows. She is becoming wilder. This, along with the fact that she is a bottle-fed calf, could end up being very dangerous for us. With everything we have been through with Butter, we have never really felt we had to be afraid of her. But whether we like it or not, we are going to have to start watching her more carefully. She is now an elk.

As winter progressed I became more and more aware of the growing distance between Butter and me. Her quirky little habits of pulling my hair and sucking my fingers were quickly becoming memories. She no longer came up to me and childishly snuggled close. This was a very sad realization for me. Through the winter I watched Butter fight for her place in the herd and for her food. She no longer ran to me for protection, and when I tried to help her or pet her through the fence she pulled away from me.

Winter Chores

It snowed and I stopped watering the herd. They prefer snow in the winter, luckily for me. To my dismay, the snow didn't last this year. A Chinook came in and all of the snow melted. I was soon back to watering the elk once again.

I had to go in for an operation and was laying

low for a while, so Carson helped out with the chores before he left for work. Of course he made me a nervous wreck when he didn't fill the oat pails to the top and used fewer pails. He told me they don't need oats every day when they had hay. I tried to tell him we had to water the elk because there was no snow, and he told me they could drink out of the puddles. So the war raged on.

I was so glad to get back on my feet. I was probably back out to the chores earlier than I should have been, but I had to get Carson away from my elk. Sometimes I think it's no wonder they try to kill him! The elk know who's is feeding them. At times, my girls call me as soon as they see me. Carson says, "Of course they make a fuss over you. You have them spoiled rotten. Why wouldn't they call for you?" I tell him they are calling for me because he doesn't feed them enough. You can't imagine his reaction! The truth is, both friends and veterinarians have told me my elk are sometimes too fat.

January finally brought the snow, and lots of it. It brought more blizzards than I had seen in many years, and the deep snow made it very difficult to move around. I had been putting oats out when the herd was in the pasture or feeding through the fence because Butter had started to grit her teeth at me. She was in a hurry to get to her oats before the rest of the herd pushed her out of the way. I had been pretty leery of her since she raised up on her legs with Carson.

On Valentines Day I went to do my regular chores. Because the herd was out in the pasture I didn't have a close view of them. Later in the day, I saw them eating their oats but could not tell who was who from the window. The next day I had my grandchildren with me all day and got out later than usual. It is not hard on the elk to miss a day, or even a couple of weeks of oats, but my daily chores are my chance to check for missing and injured animals. It was around four-thirty when I got out, and as I fed the herd I was getting uneasy. I could not see Butter anywhere. Her calf was in, but there was no sign of her. As I searched tags on the cows in case she had been pushed to the back of the herd, my heart dropped to my toes. I knew something was wrong. There was absolutely no sign of her.

Carson had stacked big, round hay bales inside of the elk fence for the last couple of years. We used to stack the bales outside the fence, but the wild deer came in and trampled and urinated on the hay. The hay became a source of food for the herd of deer, and we couldn't afford that. Sometimes I noticed different elk standing right on top of one of the bales of hay. They love to climb and paw at the bales. Carson brought the loader in every couple of weeks and packed the food on the tines of the loader, breaking the huge bales apart so the elk could feed easier. I had worried in the past about one of the elk getting stuck between the bales, but none had so far.

Butter Gets Trapped

I drove over to the big gate and parked the truck just outside. I was worried about getting the pickup stuck in the deep snow. As usual, the men were all out working. I turned the pickup off so I could hear Butter if she was calling out. I pushed the big gate back and forth, bucking the snow to make a path for the pickup in case I had to drive through. I was calling Butter's name at the same time. On the third call, I heard her answer me. Her voice seemed so far away as I stumbled my way through the snow toward her calls by the haystack. As I got closer, I could hear the urgency in her call. I kept calling her name. I knew Butter was trapped in the bales, but I didn't know where she was. I didn't know if she was head first, feet first, how far down she was, or whether or not she had broken anything while fighting to get out. I climbed to the top of the bales and finally saw her. Two feet were down in the hay and her head was still on top of the bales with her other two legs. She had no traction to push her way up. When she saw me she quit calling. Butter had fought so hard she was exhausted and lathered in sweat. It was very cold out and we were going to be lucky if she didn't end up with pneumonia.

"Butter, I'll be right back," I said, as I patted her

head. I knew I had to find ropes or chains—anything that was long enough to reach that bale and pull it out. Once it was out of the way, Butter would be able to maneuver. I returned through the drifts of snow to the truck. I backed it up all the way to the shop. The bale I had to pull was holding Butter's back end up. Once her back legs were down she would be able to push her way back up, just as long as her head didn't go down first.

I had to make sure I had enough rope and chain to wrap around the bale and then back to the truck so that when I pulled the bale out it didn't land on the pickup. I put the truck in 4x4 low and headed back to Butter, driving like a madwoman through the deep snow. Butter really needed me this time. The look on her face told me she had been giving up her fight. I spun through the snow with the truck shuddering and its wheels spinning. It felt like the truck was bucking, but I kept on moving. I drove the pickup back and forth, making a trail to the bale where Butter was stuck so that when I pulled the bail I could keep away from it. I hooked four of the longest chains together and then to the pickup, tying ropes to the other end of the chain. I had to stop to suck air into my struggling lungs. The chains were very heavy and it had proven exhausting to load and unload them, then lay them all out. I started to climb the bales, following a path the elk had trampled.

Butter hadn't moved at all. She just lay there looking at me. I kept talking to her and she sadly returned my calls. I had not heard anything so sad since she lost her twin calves. I told her, "I'll get you out—you'll see"—hoping I was right. Finally I got the rope wrapped around the bale. I had to take a few breaks to catch my breath.

Gasping, I half fell, half jumped off the stack to the ground. I crawled to the truck on my hands and knees through the snow and struggled to my feet and crawled into the pickup. I had worked up such a sweat; I took off my toque and mitts. I actually thought my cute, little, fat body was going to give up on me, but *for the love of Butter* I could not give up!

I took three or four long shuddering breaths of air and put the truck in reverse. I stepped on the gas with a vengeance. The bale came flying down toward me but I kept backing up. There was snow flying everywhere and I couldn't see. I knew the bale hadn't landed on me when I stopped. I piled out of the truck to see where Butter was, and she was just lying there. She wasn't even trying to use her back legs to push herself up. She was scared, but at least her back legs were together.

I climbed back up into the bales, grabbed her around the neck, and pulled. With a mighty lunge she heaved herself up, knocking me down. I rolled right to the bottom of the stack into the deep snow.

I just lay there, knowing I had to pick up all the rope and chains. Only I had reached the end of my rope. I looked at my watch. It was 6:30! It had taken me two hours to pull Butter out, and, in my fear, I had not even noticed it was dark out.

Butter came down off the bales and stood beside me with her head hanging down. I picked up loose hay and rubbed her body, trying to dry her off. It was minus twenty degrees and she was wet from exertion. I figured that she had been stuck for at least a day and a half. She was lucky I found her when I did. I'm sure one more day would have been the end of Butter.

A Sad Time Ahead

The end of March arrived and spring was in sight. It was time for me to start making decisions about our elk business. Although I had pulled Thunder out in the fall, Butter had unlatched the gate and he ended up back in with the cows for a couple of days before we could separate them, so there will probably be a few calves in the spring. They will be born in late May or early June.

Carson and I debated about whether to sell—that is, basically give away—my bulls. Beyond butchering them, they really had no monetary value. Once the one and only case of CWD was found in Alberta—a wasting disease that is found in the cerved family—the elk industry collapsed. I just couldn't afford the upkeep of the bull farm. The cows would also have to go. I decided I could not part with Butter and her calf, nor perhaps, with Eve, Carson's elk. They will live out their days here as our friends and pets. I could not part with Butter if my life depended on it.

I put all of my life and love into my beautiful animals, but the obstacles have become too great. For seven years, I have looked forward to spring and the new life it brings, but this year I only saw sadness ahead for the animals and me.

I had a young man coming to pick up the yearling bulls and take them to the bull farm. He phoned to say he was on his way, with no advance warning and no men around. I figured that if I could get them on the runway and swing the gate shut that he and I could separate them without any help. There are two gates to close once the herd is in the runway. The first separates them from the big area so they can't run all over. The second is about five hundred to six hundred feet and locks them into the facility. This is the gate the men have to run to slam shut. It is dangerous because the elk are running in fear and could run right over you. The only time the facility is used is for needling, separating, and for pulling a calf.

Butter's Heroic Rescue

Somehow I had lucked out and the herd had all run down the runway. Butter, of course, had not run with the herd and squeezed in with me before I could close the gate. I was leery she would toss her head and prance around, so I stood still for a few minutes watching the herd at the other end of the runway. If the herd saw me, I thought they would turn and run into the facility. That would give me enough time to run the gauntlet and close the second gate. A few cows might get by me, but most of them would get locked in. I guess I should have quit while I was ahead. I took a deep breath and started to run with Butter loping along next to me. She had been ornery with me all winter, so I was a little nervous that she might hurt me. As I ran I talked to Butter to see if it would calm her. "You could be down there in five seconds. I should have taught you how to shut a gate. You know how to open gates—why not close them." As I ran toward the herd, they turned and ran back into the facility. I made the gate and was hollering as some elk turned to run toward me. I waved my arms and pulled on the big gate. The darn thing was caught on a lump of dirt! I was trying to lift it over the dirt when the first cow hit me. I went sailing backward into the boards, falling to the ground.

Butter skidded to a stop. Another cow hit her and spun her around. She almost went down, but she managed to hold her feet and crossed in front of the running herd to get to me as the rest of the herd came down on us. Butter stepped over me as another cow hit her. She stepped sideways onto my leg. When she felt my leg under her hoof she tried to pull her weight onto her other legs and stumbled again. Regaining her balance, she started to bark. Elk do this when calves or the herd are in danger. That is the only time I have ever heard them bark—and it made my hair stand on end.

Butter was mad. She went up on her hind legs, striking out at any cows and calves that were getting close to us. Somehow she remembered I was under her feet through the whole ordeal. Even when Baby Butter came too close, she was pushed away by Butter. I watched the first ten or fifteen elk run past Butter and me. I finally stopped screaming and tried to help Butter. The more I screamed, the crazier she got. I rolled myself into a tight little ball and covered my head with my arms, rolling as close to the fence as I could get. Elk are like horses: if they can see you they'll do anything they can to avoid stepping on you.

Aside from Butter stepping on my leg, I was not hurt. The herd went thundering by, back toward the gate they had come in. They still could not get out. They were all milling around as I pushed myself up off the ground. I staggered to the opening where I could squeeze through to get out of the runway. Butter squeezed her big body through the opening after me, mewing as she came, as though she was trying to ask me if I were okay. I leaned against the fence crying as the shock of what had happened started to set in.

Bev—Now in Training to Be an Elk

As I stood there, Darrel Hunter (from the bull farm) and his little boy Keaton came running over to me. "Bev, why didn't you wait for me? I saw what happened. I thought you were dead for sure. You sure have one heck of an animal there. If I hadn't seen it I wouldn't believe it." I had told Darrell a long time ago that if anything ever happened to me he was to take Butter. I'm thrilled by his love for elk and I thank God every day for finding him. I remembered when our friend Dwayne Veldhouse had introduced us to

Darrel and his family. I was never so happy to see anyone in my life as I stumbled, crying, into his arms.

Butter was back to prancing around Darrel and Keaton and was challenging them while she tried to decide if they were friend or foe. We left the pasture to let her have some time to settle down. As we left she hurried over to the fence to touch her calf's nose.

Butter—My Lifelong Friend

It's funny how wrong a person can be. I was afraid of the animal I raised, yet she probably just saved my life. I guess when she grits her teeth and pulls away from me she has her reasons, but it has nothing to do with her love for me. She is an elk and that's how they treat each other. All she is trying to do is teach me how to be an elk. After all, I raised her and treated her like a person!

One thing I do know is that, *for the love of Butter*, I will always do things for this animal, even things that some people won't do for another human being. At the same time, for the love of Beverly, Butter has done things that are never expected from an elk. I have many friends, but none like my friend Butter.

I plan to take a break from writing about Butter for a while, but will always keep a diary about her life. Who knows, perhaps in a couple of years enough will happen that I will need to write another book.

After all, trouble seems to follow her, and I can't imagine that ever changing.

If you ever want to meet Butter, just come to Seventh Ave Elk Ranch in Manning, Alberta. She will welcome you with open hooves!

BEVERLY LEIN was born in Manning, Alberta, and raised in nearby Sunny Valley on her father's farm with her three sisters and two brothers. Her hobbies have always included writing stories, songs, and poems. Before raising elk with her husband Carson on their own ranch, Lein worked as a cook, a farmer, a retailer, and a small business owner. A proud mother of two and grandmother of five, Lein's life also requires her to be a veterinarian, a midwife, and an elk counsellor—depending on what the day brings.